The Practical Handbook of
CARPENTRY

By R. J. DeCristoforo

Fawcett Publications, Inc.
1515 Broadway
New York, New York 10036

FRANK BOWERS: *Editor-in-Chief*

SILVIO LEMBO: *Creative Director*

HAROLD E. PRICE: *Associate Director* • HERB JONAS: *Assistant Director*

JOSEPH C. PENELL: *Marketing Director*

RAY GILL: *Editor*

LUCILLE DE SANTIS: *Production Editor*
JULIA BETZ : *Assistant*

Editorial Staff: DAN BLUE, ELLENE SAUNDERS, LARRY H. WERLINE

Art Staff: MIKE GAYNOR, ALEX SANTIAGO, JOHN SELVAGGIO,
JOHN CERVASIO, JACK LA CERRA, MIKE MONTE

How-To Art by Henry Clark
Cover Color by John Capotosto

Printed in U.S.A. by
FAWCETT PRINTING CORPORATION
Rockville, Maryland

FOURTH PRINTING

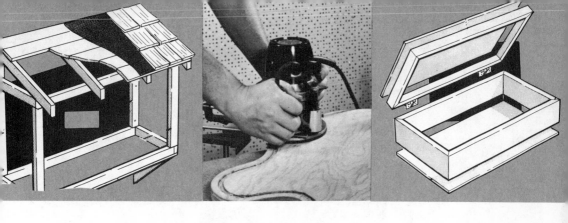

CONTENTS

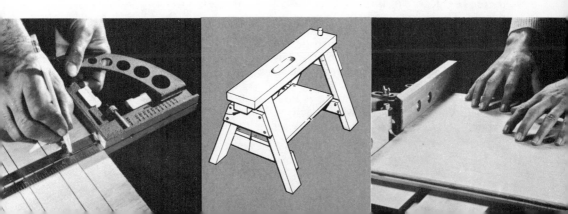

The home workshop, in addition to being a wise investment, can be a place for family activity. Youngsters enjoy woodworking, wives love the furniture projects that result.

YOUR HOME WORKSHOP

Start small and build your shop to suit your budget and your needs

It's the man who makes the shop. Great projects and a lot of home maintenance can come from a shop that is nothing but a drawer used to store hand tools. But it would be foolish to deny that similar work comes easier when you are set up and equipped with tools that help get you there. The degree of this aid can confuse people. The dream workshop is nice but isn't necessary. On the other hand there is a minimal you should try to get to pretty fast if only to avoid frustrations that can bug you enough to send you to TV or knitting.

· As far as workshop furniture is concerned, you do want a workbench—a cabinet above it or near it—a couple of sawhorses—and you pretty much have it made. You can go further, *much* further, and if you get hep to the good things that can come out of having your own workshop, you probably will. Getting to the ultimate can come in easy stages but you can be a very efficient craftsman long before you arrive at that point.

THE LOCATION

Where you put the workshop depends on your own situation. Two common places are the garage or the basement. I prefer the garage because it's usually easier to get in and out of with supplies and equipment. On the other hand it's

4

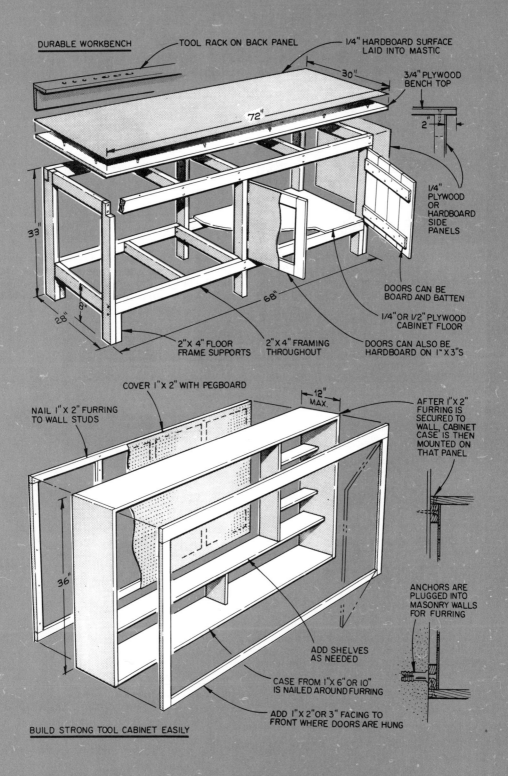

DURABLE WORKBENCH

TOOL RACK ON BACK PANEL

1/4" HARDBOARD SURFACE LAID INTO MASTIC

30"

3/4" PLYWOOD BENCH TOP

72"

2"

33"

8"

28"

68"

1/4" PLYWOOD OR HARDBOARD SIDE PANELS

DOORS CAN BE BOARD AND BATTEN

1/4" OR 1/2" PLYWOOD CABINET FLOOR

DOORS CAN ALSO BE HARDBOARD ON 1"X3"S

2"X 4" FLOOR FRAME SUPPORTS

2"X 4" FRAMING THROUGHOUT

COVER 1"X 2" WITH PEGBOARD

NAIL 1"X 2" FURRING TO WALL STUDS

12" MAX.

AFTER 1"X 2" FURRING IS SECURED TO WALL, CABINET CASE IS THEN MOUNTED ON THAT PANEL

36"

ANCHORS ARE PLUGGED INTO MASONRY WALLS FOR FURRING

ADD SHELVES AS NEEDED

CASE FROM 1"X 6" OR 10" IS NAILED AROUND FURRING

ADD 1"X 2" OR 3" FACING TO FRONT WHERE DOORS ARE HUNG

BUILD STRONG TOOL CABINET EASILY

5

A radial arm saw is the big power tool in this shop. Getting power tool first makes it easier to build the shop. Workshop furniture can be plain or fancy, to your taste.

A pair of sawhorses make sense regardless of the kind or size of your workshop. The handy hinges shown make sturdy, portable workbench area possible—easily stowed.

Even little things like Allen wrenches and chuck keys will take up small space and be readily available if you plan for storage. Design a small shelf with holes for items.

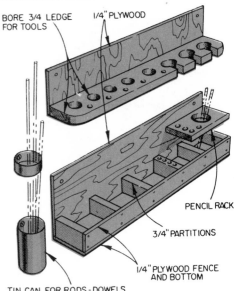

TOOL RACKS FOR DOOR BACKS

BORE 3/4 LEDGE FOR TOOLS

1/4" PLYWOOD

PENCIL RACK

3/4" PARTITIONS

1/4" PLYWOOD FENCE AND BOTTOM

TIN CAN FOR RODS - DOWELS

usually easier to heat a basement. But I've seen workshops in attics and even in a closet in an apartment.

It's nice to have a reasonable amount of space but you can put an efficient radial arm saw shop in a narrow area which you often find on the side of a one or two-car garage. You can equip yourself with an impressive array of portable power tools and need but a closet to

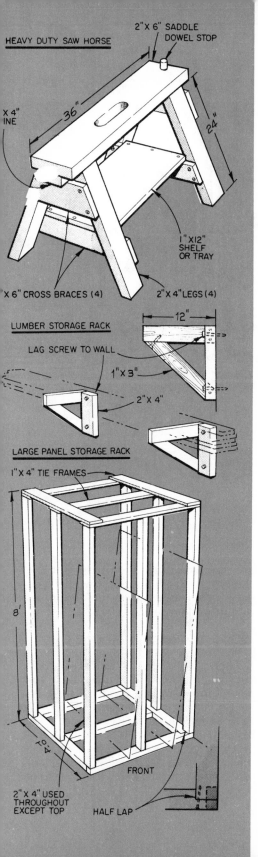

HEAVY DUTY SAW HORSE

2"x 6" SADDLE
DOWEL STOP

X 4"
INE

36"

24"

1"x12"
SHELF
OR TRAY

x 6" CROSS BRACES (4)

2"x 4"LEGS (4)

LUMBER STORAGE RACK

12"

LAG SCREW TO WALL

1"x 3"

2"x 4"

LARGE PANEL STORAGE RACK

1"x 4" TIE FRAMES

8'

2"x 4"
4"

FRONT

2"x 4" USED
THROUGHOUT
EXCEPT TOP

HALF LAP

house them. Floor space becomes important when you think in terms of individual power tools set to stay. Each tool requires room for itself

If it's possible to set aside an area exclusively for the workshop, do some preplanning before you start building. Check through the sections on tools so you will have some idea of the equipment you will start with and what you may add. Also, you can make some wise decisions relative to the space you have to work with.

By way of example, let's describe what we feel would be a real nice setup for a homeworkshop. It would consist of the workbench and cabinet shown in the drawings which, incidentally, can be built with just a handsaw and a hammer. A couple of sawhorses for use near the bench or anywhere in or about the house. A table saw sitting near the bench area. Some hand tools and portable power tools that can be stored in the cabinet or under the workbench.

EXPENDITURES

You can spend as much as $500.00 for a table saw but excellent ones are available for $100.00 to $200.00. For about $100.00 you can get three or four portable power tools. For another $30.00 or $40.00 you can get a supply of hand tools. Figure another $30.00 or so for materials you need for the workshop furniture, and you can see that the cost of setting up for *efficient* homeworkshop efforts really isn't all that bad. You can easily save what you spent in short order by using the tools to do house jobs that would otherwise require expensive professional help.

Saving money by doing things yourself is not a sales pitch by any means, not these days when you can figure that 50% to 60% of the cost of any job is in the labor. The money you put into tools is an investment not just in an economic sense but in your own well being. Being able to move out to the shop to get away from your money-making job, the TV set and the newspaper headlines can help you live happier and probably longer.

Professionals often assemble vertical framing on the ground, then tilt it into place. Good when working with others, otherwise try our one-man plan outlined in the text.

BASIC HOUSE FRAMING

Whether you build a house yourself or not, it's good to know how

Chances are you've seen a house under construction and possibly the array and arrangement of structural members planted the thought that this might be a bit too much for you to tackle. Not so, really; the rough framing is the easiest part of building. It can be, and often is, a one-man job when tackled by the do-it-yourselfer. It's not likely that the average person will ever hand-build his own home. It's possible you may never even add a room or do similar king-size jobs; but even so, all homeowners should know what's hidden in the walls and how it was put there. This makes it easy to locate structural pieces when you

are constructing a built-in, placing shelves, adding a door or a window, doing a paneling job, even hanging a picture or curtain rods.

FLOOR JOISTS

There are some differences in house construction under the first floor. This has to do mostly with whether there is a basement or a crawl space. But above that the framing is pretty routine. If you check the sketches that illustrate the two types of construction you will see that one way or another the perimeter walls of the house put their weight on the foun-

Often used over basement-type foundations are 2×12 joists. Blocking is cut from the same material. Sub-floor or 1×6 boards or plywood. Good for over a crawl space too.

Headers are usually cut from uniform 4×12 material today, regardless of span. This costs more in material but saves on labor. The filler studs are yet to be added here.

"Let-in" bracing is fitted in notches cut in all members the brace crosses. Outline cuts done with saw, waste chiseled away. Notch should provide tight fit for brace.

Windows fit in the rough opening left in the rough framing; be sure of window size before starting work. Note "bird's-mouth" cut in rafter coming over the top plate.

dation. They may sit on a sill that has been floated in the concrete immediately after it was poured, but this is true mostly of houses built on concrete slabs. In the two examples shown, the bottom plates on which the wall is built are nailed through a sub-flooring into members that are attached to the sill.

Sturdy 2×12s are used to span perimeter walls when the house has a basement. This can also be done over a crawl space, but more often, as the sketch

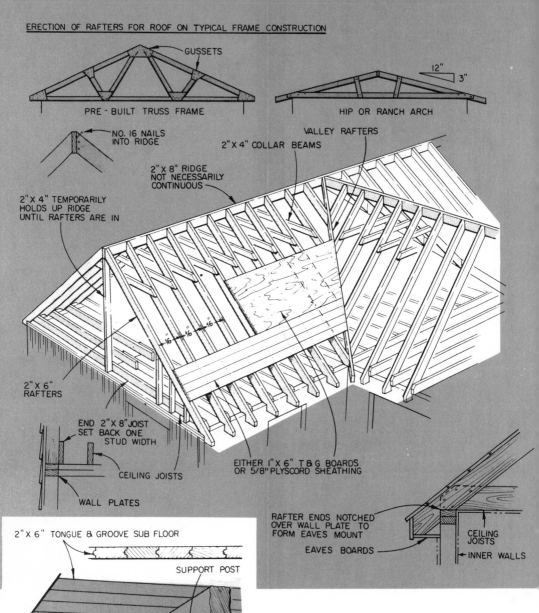

GUSSETS

PRE-BUILT TRUSS FRAME

HIP OR RANCH ARCH

12"

3"

NO. 16 NAILS
INTO RIDGE

VALLEY RAFTERS

2" X 4" COLLAR BEAMS

2" X 8" RIDGE
NOT NECESSARILY
CONTINUOUS

2" X 4" TEMPORARILY
HOLDS UP RIDGE
UNTIL RAFTERS ARE IN

2" X 6"
RAFTERS

16   16   16

END 2" X 8" JOIST
SET BACK ONE
STUD WIDTH

CEILING JOISTS

WALL PLATES

EITHER 1" X 6" T & G BOARDS
OR 5/8" PLYSCORD SHEATHING

RAFTER ENDS NOTCHED
OVER WALL PLATE TO
FORM EAVES MOUNT

EAVES BOARDS

CEILING
JOISTS

INNER WALLS

2" X 6" TONGUE & GROOVE SUB FLOOR

SUPPORT POST

SILL AND
ANCHOR BOLTS

CONCRETE
PIER

4" X 6" GIRDERS ON
4 FT. CENTERS (ALTERNATE
TO 2" X 8" ON 16" CENTERS)

FLOOR FRAMING OVER CRAWL SPACE

shows, a post-and-beam type of under-structure is used. On post-and-beam work, it's common to use 2×6 T&G boards of 1-1/8″ thick plywood as the subfloor. Usually, over 2×12s, 1×6 boards are used layed down in a diagonal pattern, but even here, plywood sheets are becoming more common.

A good one-man procedure for construction above that point—that I have used successfully on several occasions—is as follows.

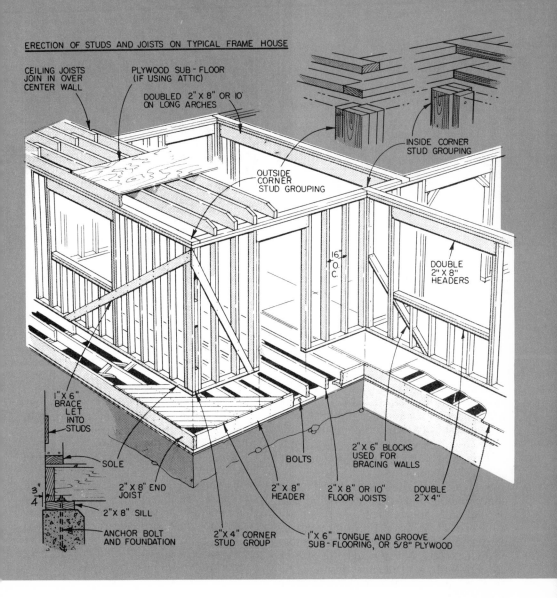

ERECTION OF STUDS AND JOISTS ON TYPICAL FRAME HOUSE

CEILING JOISTS
JOIN IN OVER
CENTER WALL

PLYWOOD SUB-FLOOR
(IF USING ATTIC)

DOUBLED 2"X 8" OR 10'
ON LONG ARCHES

INSIDE CORNER
STUD GROUPING

OUTSIDE
CORNER
STUD GROUPING

16"
O.
C.

DOUBLE
2" X 8"
HEADERS

1"X 6"
BRACE
LET
INTO
STUDS

SOLE

3"
4"

2" X 8" END
JOIST

2"X 8" SILL

ANCHOR BOLT
AND FOUNDATION

2"X 4" CORNER
STUD GROUP

BOLTS

2" X 8"
HEADER

2"X 8" OR 10"
FLOOR JOISTS

1"X 6" TONGUE AND GROOVE
SUB-FLOORING, OR 5/8" PLYWOOD

2" X 6" BLOCKS
USED FOR
BRACING WALLS

DOUBLE
2"X 4"

CORNER POSTS

Check the sketch that tells you how to assemble corner and partition posts and make as many of these as you need. For these sub-assemblies, select some nice, straight studs. Usually the procedure is to nail spacers to one stud, then add the second stud and then the third that forms the angle. Use 10d nails throughout and stagger them; 12" spacing is about average.

Set up your posts at correct locations using temporary bracing to keep them in position. Be sure to use a level on two adjacent faces and that the bracing is strong enough to keep the posts in vertical alignment. Attach to the bottom plate with two 8d nails on each face. These nails are driven in at an angle of about 45° as you brace the opposite side with your foot. Even then, the post may shift a bit but you can adjust for this when you drive the other set of nails.

Careful workmanship on rough framing will make other jobs easier. Check with a level and a square as you go along. These walls are not yet ready for finishing materials.

Items similar to the metal brackets being used on the above rough framing are today available for the assembly of most of the units in house-skeleton construction jobs.

WINDOWS AND DOORS

Next step is to mark window and door locations on the bottom plate. Cut headers to correct length and then make a sub-assembly by attaching outline studs in place with 10d nails. You can, on this sub-assembly, add the filler studs or you can install them later.

Set up the header assemblies using some bracing and again toe-nailing with two 8d nails on each face. At this point you probably have enough vertical pieces up to support the bottom member of the top plate. Nail through this at each crossing with two 16d nails, driving extra ones as you cross the headers. Use the level frequently as you go, on all vertical pieces being nailed.

INSTALLING STUDS

When this is done you can start adding studs between openings. These are spaced 16" on centers, and started from corners toward window and door openings. To do the job right, and to make it easier, cut a spacer from a piece of 2×4 so you can use it to locate studs accu-

rately. The length of the spacer should be 16" minus the full thickness of one stud. This not only spaces the studs but acts as a brace when you are toe-nailing. Do the usual 8d toe-nailing at the bottom end of the studs; drive 16d nails down through the plate at the top.

After all studs are in, you add the top member of the upper plate, using 10d nails staggered about 16" apart. Be sure to check the illustrations that show the lap joints that occur over corner and partition posts. When plate members must be spliced to get adequate length, let the joint occur over a stud. This is especially important with the bottom piece.

SILLS AND FILLERS

If filler studs were not set in before, place them now—nailing to outline studs with 10d nails spaced and staggered about 16". Add the sills and short studs at the window openings, nailing through the sill at each stud crossing with two 10d nails. Attach the top part of the sill with 10d nails spaced and staggered about 8". Both sill pieces, as you add them, are toe-nailed at each end into the filler studs.

Consider side-wall insulation at the same time you do the framing. Bats are stapled into place between the studs. An easy job now, but hard after the walls are covered.

Lengths of 2×4 can be fitted between studs as bracing or you can "let-in" 1″ stock. Notches for the 1″ stock should not be cut carelessly—the better the fit, the more strength the brace will have.

If you have done a good job installing studs, you can pre-cut as much fire-blocking as you need. The pieces may be installed on a line or they may be staggered as shown. Staggering does make it a bit easier to do the nailing.

The word "rough" in rough framing should not be taken as an excuse for careless work, especially as far as alignment is concerned. Use the level frequently as you go, and a square if you need it. You'll appreciate a careful job here later on when the time comes to install doors, window and wall coverings.

CEILING JOISTS

What happens above the structure you now have starts with joists that span the perimeter walls. Their size will depend on the load they must carry and specifications in the local building code. This is something you must check into anyway before starting the kind of job we have been discussing. Many people are hesi-

tant about doing this and they can't make a bigger mistake. Local codes are established for your protection and in addition to keeping you in line so you do a *safe* job, they provide a tremendous amount of information—even to telling you correct size nails and how many to use. These codes can vary from area to area if for no other reason than different weather conditions. For example, heavy snow falls mean a roof must carry greater loads than a structure in an area that gets nothing but rain. Cooperating with building inspectors does pay dividends.

ROOF FRAMING

Doing the roof is a bit harder than vertical walls but only because angular cuts, usually, are involved. A shed roof is no problem but a hip roof with gables is something else again. However, the angles are the same for all similar pieces regardless of length. Once you have established it—by trial-and-error if necessary—it becomes the template for making duplicate cuts. If you are working with a radial arm saw or a portable cut-off saw, the same tool setting is used on the tenth piece as was on the first.

Basic roof framing can be pre-assembled pieces like those shown in the drawings. To make these, you can actually use the sub-floor as a platform. These will be heavy, and it's one time when a couple of strong sons come in handy—or a kind neighbor?

If you don't pre-assemble, put the ridge up first, using temporary supports to keep it there. Working from this point to a wall, it is not a hard task to cut one rafter correctly and then use it as a pattern to cut the others. Where you have gables, start with ridge pieces, add the valley rafters and then work on from there. In the example we show, the ridge cut on all the common rafters is exactly the same and so is the cut where the rafter passes over the outside wall. Make cardboard templates of these cuts—thus you can use them as patterns regardless of rafter length. When the ridge cut differs, you can still use the plate-cut template.

Ten-inch table saw has a built-in motor and a stand to which you can attach casters. It is therefore not too difficult to move it about in a shop with limited floor space.

TOOLS FOR SAWING

Buy the tool that suits the kind of projects you have in mind to do

A table saw is a very fine tool to have in any woodworking shop. If you can't have one right off, it would be unwise not to plan on getting one as soon as possible.

The above statement might very well be made concerning any of the following sawing tools:

Radial arm saw
Bandsaw

Jigsaw
Portable cut-off saw
Saber saw
Bayonet saw
Some hand saws

The truth is that almost any power tool you acquire is going to help you do better, faster work. Few of us can afford them all and many of us don't require the entire parcel. If you have a good hand-tool shop, don't do too much in the way of furniture projects; but need a fence or some rough planters, maybe want to install some headers or some grid work for concrete, a good choice for you would be a portable cut-off saw. This does not imply that such a tool is limited to work in those areas, merely that it is a good choice under the conditions described above.

The point is that any saw will saw any wood but the design of each makes it especially useful for particular applications. So check through the descriptions that follow and mate the tool with the jobs on hand or with your thoughts about how far you want to go as a homework-shop craftsman and in what areas. Then make a choice—at least for the *first* one.

THE TABLE SAW

The table saw has its blade under the table with built-in adjustments so the blade may be raised, lowered or tilted. Work is placed on the table and moved against the blade; with a miter gauge for crosscutting, or against a fence for ripping. In either situation you can make angular cuts by working with the blade tilted. Thus, in addition to square cuts, you can form miters, bevels, chamfers

"Compact" versions are very popular these days. Though small (this one has a 7½ in. blade) they do all that the bigger models do. Some other cuts have capacity limits.

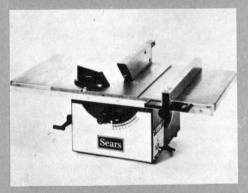

Controls are used for raising, lowering or tilting the blade. Tilting a table saw—you tilt the table instead of the blade for angular cuts—is practically passe today.

The miter gauge is used to move the work against the blade for crosscutting. Refer to instructions in tool manual for maintaining alignment of your tool components.

Don't use rip fences as a gauge to cut duplicate pieces. Instead, clamp a piece of wood to the fence as a gauge to determine the precise length of the cut-off point.

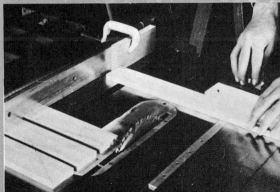

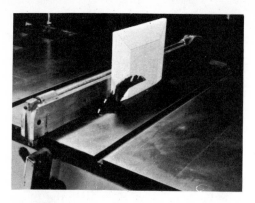

Tilt blade 10° or 15°, pass work through on edge to "raise" a panel. Use piece of wood to pass work through for small work. This is also the way to bevel and chamfer.

A dado makes rabbet cutting a snap. One pass does what takes many with a regular saw blade. You need special table inserts for dadoing tools and for molding heads.

Good idea to collect blades. The blade on the machine is for crosscutting and ripping. Special blades for ripping (above), super-smooth cuts on lumber available too.

The blade on the radial arm saw is above the table. It can be raised, lowered or tilted. In order to set it for miter cuts it is necessary to pivot the overhead arm.

or compound angles. Any of the basic saw cuts can be accomplished on a table saw by working with just the saw blade that comes with the machine. A groove or a dado wider than the saw kerf is accomplished merely by making repeat passes. A rabbet can be done the same way or by making two cuts that leave an L-shape in the wood. If you follow this same procedure on four sides at the end of a piece of stock, you end up with a tenon. By cutting with the blade tilted and the stock on edge, you can "raise" a panel.

THE DADO ASSEMBLY

To do some of these jobs faster, buy a dado assembly or an adjustable, one-piece equivalent. A 3/4" groove that would require six or seven passes with a regular saw blade can be done in one pass with a dado. Think of the dado as a saw blade that can be adjusted for the width of cut you want.

THE MOLDING HEAD

With a molding head (another acces-

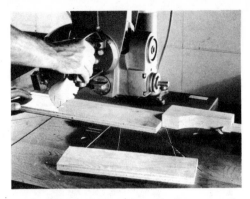

To crosscut, hold the work down on the table and pull the blade through. Block over the fence is a stop for gauging the length of your other duplicate pieces.

Dado used on radial arm in normal cross-cutting position or as shown, parallel to table so long work can be moved across it. Dado protrudes through the homemade fence.

Molding head may be mounted in place of a saw blade. Some models have special guards to mount over tool of this kind. Above— equally spaced cuts cause incised scallops.

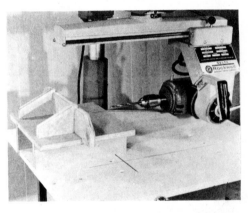

Modern radial arm saw will take a chuck to grip drill bits and other tools. Drill table shown above is homemade and locked in the table in place of a regular fence.

sory) you can shape decorative edges or do more practical things like forming glue-joints, cabinet-door lips, drop-leaf table joints, and so on. The more knives you buy for the molding head, the more jobs you can do with it. But get them as you need them—don't attempt to buy a variety just to have them on hand.

THE RADIAL ARM SAW

The radial arm saw has its blade above the table, locked to a motor that rides an overhead arm. Work, for crosscutting operations anyway, is placed on the table and held stationary while you pull the saw blade through to make the cut. This can be a great asset when you need to, for example, trim the end of a long 2×4. Many craftsmen feel the concept leads to easier accuracy when doing miters and compound-angle cuts. For ripping operations the blade is locked in a position parallel to the fence and the work is moved against it.

Because it provides convenience in handling long stock, the radial arm saw is the tool you see at any house-construc-

A similar table makes it possible to use a radial arm saw as an efficient disc sander. Most catalogs list discs as accessories but you will have to make the table on your own.

Typical of other tools you can mount is a rotary planer. In addition to doing surfacing jobs it can be set up for decorative work similar to that as shown above.

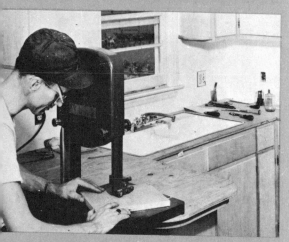

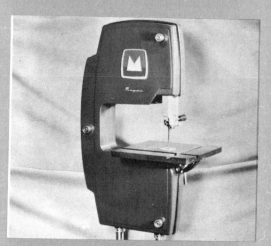

None of the homeworkshop bandsaws are so big that they cannot be carried for at-the-job use as this kitchen-remodeling craftsman in the above photo discovered.

Using power tool acquaints you with accessories you can make yourself—for making jobs easier and even possible. V-block is a carrier and length gauge to cut dowels.

This typical homeworkshop type bandsaw shown above is a Magna American product. It can be used either as an individual tool, or for mounting on the Shopsmith.

In this shop, bandsaw comes with a special speed control and special blades for cutting metal. You can cut non-ferrous metals with an "all-purpose" blade, but not steel.

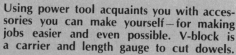

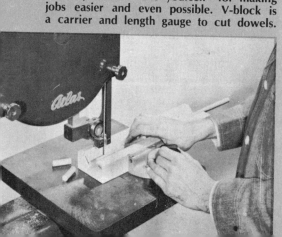

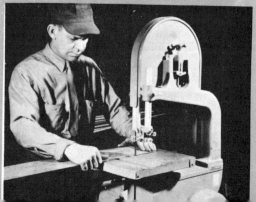

tion site. At one time it was used almost exclusively in such applications. But that was before it was "re-designed" to be a good tool for the homecraftsman. Today, it is so made and there are so many accessories available for it, that it is often called a one-tool shop.

SPECIAL ACCESSORIES

In addition to the dado and the molding head jobs we mentioned for the table saw, you can use the radial arm to do quite a bit of drilling, sanding, grinding, polishing, surface planing, shaping, drum sanding. For some models you can buy special tool accessories to be powered by the saw motor. A saber saw attachment is one example, a small lathe is another. On one model called the "Saw-Smith", you can actually mount separate tools like a bandsaw, belt sander or jointer. The machine has a variable speed control to give the right speeds.

As you can see, as far as accessories are concerned, there are many more available for the radial arm saw than for the table saw. Both are stationary tools but either can be moved to a construction site or to a room in the house where you are doing considerable cabinet work.

THE BANDSAW

The bandsaw runs an endless blade over two wheels and can, with a narrow blade, turn a ¼" corner or, with a wide blade, cut through a 6×6 with ease. It is the fastest woodcutting machine and, between the two examples cited above, are a host of other jobs you can do. But it's not likely to be the first wood-sawing tool you will buy.

Even though its depth-of-cut is great, its width-of cut is limited by the distance from the blade to the throat—11 to 12" in a typical home workshop machine. Slicing a plywood panel down the center either way would be impossible, but if you needed to cut a cabriole leg from a piece of 4×4 stock, the bandsaw is the only power tool you could do it on. If you wish to cut 1" boards from a piece of 6" thick material—you can do that on a bandsaw. If you needed 12 duplicate pieces in ¼" stock, you could stack them and cut them all at once on a bandsaw. You can't form grooves or cut rabbets on long edges and most bandsaw cuts require more attention for smoothing than cuts made on a table saw or a radial arm saw.

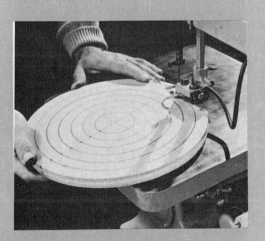

You can do internal cutting with jigsaw without a lead-in cut. It's possible because you can slip the blade through hole and *then* chuck the blade into the machine.

It is also possible to do fancy fret-work by the process of "piercing." It's really as simple as drilling through each of the areas of the stock that must be cut out.

Heavier jigsaw blades should be gripped in the lower chuck so that they can operate like a saber-saw blade. Smaller blades are really too flexible to be used in this way.

An accessory like "Sure Square" is good to consider for making accurate cuts. Gripping the handle clamps the accessory to the work and provides edge saw base can ride.

Jigsaws are really great for getting kids truly interested in woodworking projects. And just a few minutes of instruction by you will get him started on a safe footing.

THE JIGSAW

The jigsaw *is not* a small version of the bandsaw. It uses short, straight blades that move in small vertical strokes; its depth of cut is usually about 2″ and it can take blades as fine as those used in a jeweler's saw. Since the blades are straight and gripped at both ends, you can insert them through a hole drilled in the work before chucking them in the machine. This permits internal cutting without a lead-in cut. It's called "piercing" and it's a jigsaw exclusive that permits fine fretwork projects.

The jigsaw *is*, as many people think, a fun tool, but since it can mount fairly heavy blades with coarse teeth, making straight or curved cuts in 2″ stock is not out of line and certainly brings it into the realm of other power tools. You can make a cabriole leg as you did on the bandsaw but here you would be limited to stock 2″ thick. You can use the jigsaw to form scalloped edges on a cornice, make a circular cut-out for a speaker, make cuts for intricate inlay work includ-

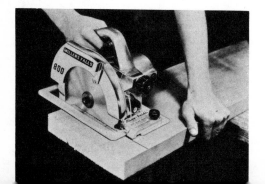

Portable cutoff saw is king when it comes to trimming work right on the job, but keep it cutting straight — and do keep the blade sharp — to avoid any chance of its binding.

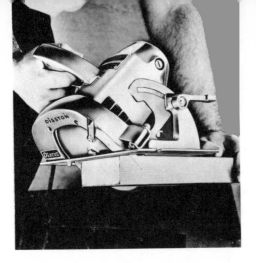

Be very sure that the model which interests you can cut through 2″ stock at 45° angle. This indicates a fairly good capacity for most of your home workshop chores.

The cutoff saw can also be used for ripping — together with the combination blade that most saws today come with or with a special rip blade which you can buy extra.

ing inlayed picture, even do pad sawing —like the bandsaw—if you keep the stack within the capacities of the machine.

THE PORTABLE CUT-OFF SAW

The portable cut-off saw is the power-saw you carry to the job—to make cuts at the site when you are building a fence, to size materials for headers and small walls, to trim wall panels, to make a cut-out in a wall and so on. It's mostly called a "cut-off" saw but it's as efficient for ripping and for angular cuts as it is for crossgrain work.

The fact that it's hand-held and easily carried about doesn't mean you can't use it in the shop for sizing-cuts on plywood and lumber and similar materials. Such a saw, backed with a good set of hand tools, makes for some very efficient home maintenance and quite a bit of project production.

It's not the tool, though, for the kind of joint-cutting you can accomplish on a table saw or a radial arm saw even

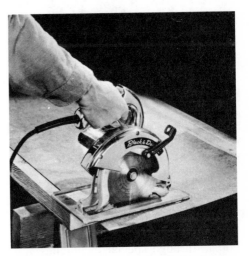

Other special blades are available for cutting metals, masonry materials, etc. You should acquire a pair of industrial safety glasses to wear when cutting such material.

The saber saw cuts its own starting hole. After you finish turning on the tool you slowly lower the blade to make contact— then continue the cutting until it pierces.

Saber saw easily follows curved lines for scallop design on the cornice board above. Forcing the cut is a big mistake — you will burn the work or perhaps break the blade.

For straight cuts, like trimming this wall paneling, clamp straight piece of wood in place to serve as guide. You can cut without it but this way is much more accurate.

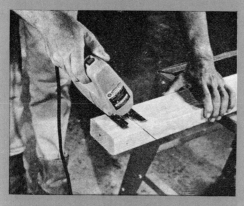

Most saber saws have a tilting base — this so you are able to make angular cuts, as shown above. Start cuts slowly, and feed slowly or the blade may wander in the cut.

Small-tooth, wave set blades are available. Using them in the saber saw provides you with a powered hacksaw capability. Metals to be cut must be firmly gripped in vise.

though, through special techniques you can use it to form grooves and rabbets. With special blades, it will cut some metal and many masonry materials.

The biggest you can afford is not the wisest choice. Big ones can be heavy and will tire you fast. A good minimum capacity standard is for the blade to be able to cut through a 2×4 at a 45° bevel-angle.

THE SABER SAW

The saber saw is a very popular tool these days. It combines many of the functions of the bandsaw and the jigsaw in a package you can hold in one hand. It can also do straight cutting but not with the speed of the portable cut-off. The blade has an up-and-down action similar to the jigsaw. Since the blade is free at one end, the tool is excellent for internal cut-outs—making switch or outlet openings in wall paneling, holes through flooring for pipes, circular cut-outs for speakers, openings through wall coverings, and so on. It can't rival the jigsaw in the fretwork area but only because it not designed to grip the very fine blades the job requires. The blades are gripped at one end only and they must be stiff so that puts a limit on how narrow and thin they can be.

A good, first handsaw is one 24" long with 10 teeth per inch. It's basically a cross-cut saw (you'll probably use it mostly for this) but it can be used for some ripping.

A keyhole saw is a good addition for making openings in walls or panels. But also buy "nest" of some blades. This will include the keyhole and some other types.

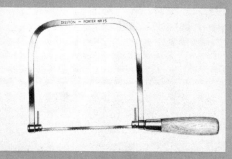

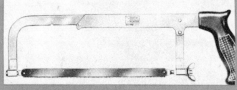

Coping saw will do some of the jobs we described for jigsaw. Since blade can be put through a hole in the work before being inserted in frame, it can do piercing jobs.

The familiar hacksaw is for metal cutting. Use it to cut bar stock, pipe and the like. Special tungsten carbide blades (they look more like rough wire) will even cut glass.

So many blades are available for the tool that it's difficult to think of a material it can't cut; wood, plastics, fiberglas, hardboards, steel, non-ferrous metals, pipe. Select the right blade and the tool will do the job. There is even a knife blade available so you can cut leather or cardboard or similar materials.

HANDSAWS

There are many types of handsaws on the market but if you are starting out and are taking the hand tool route, the ones to be concerned with are those examples we show in the photographs.

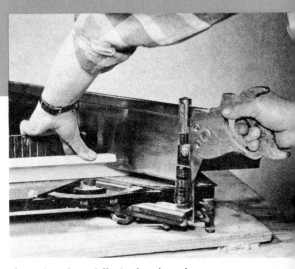

The miter box falls in hand tool category. It uses a backsaw that can lock accurately for various angles. Good for making picture frames or when doing work on molding.

23

TOOLS FOR DRILLING

These are versatile tools and can be used for many purposes.

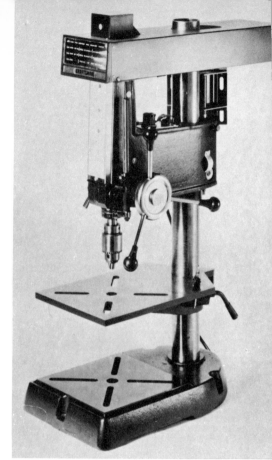

Typical home workshop-type drill press you can situate on an available counter or on a stand you buy or make. This Sears model has a built-in light to illuminate the work.

The drill press, because of the many practical accessories available for it, is one of the more versatile pieces of stationary power equipment. Few people buy it anymore just to drill holes—routing, shaping, some types of sanding, surface planing and mortising are among the woodworking chores you can accomplish professionally on a drill press. Actually, it should probably be number two on any list of tools to acquire—to come after you have made a decision on what you are going to saw with.

Drill press capacity is twice the distance from the column to the spindle center. Thus, a 15″ model lets you drill holes in the center of a board 15″ wide. The distance from the chuck to the table varies a bit from make to make but in all cases increases substantially from bench model to floor model. A choice could be based on whether you have more floor space than bench space—the difference in cost between the two types is not all that much. On the floor model you are paying for a longer column and the base; the rest of the machine is the same.

The right speed for the job is quite important so the drill press is equipped with step pulleys to provide about four of the most-used speeds. A more ideal solution is a variable speed mechanism or—and it's possible these days—equip the tool with a variable speed motor. Such an item plus step pulleys can give you a fantastic range of speeds. Do, however, check out the top speed for the tool you own or plan to buy. Drill-press spindles are not designed to stand up under the rpms of a shaper.

PORTABLE ELECTRIC DRILLS

Portable electric drills are more versatile today than ever. This is due to the fact that so many models these days have built-in speed controls. This makes it possible to work efficiently and safely in various materials with just one drill whereas, not too long ago, at least two

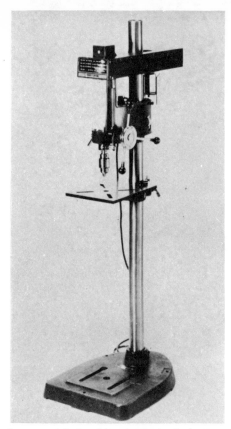

Floor model, in this photo, is the same as first shown but it has a longer column and its own base. Either of these tools can be equipped with a Sears variable speed motor.

All drill presses are equipped with stop-nuts to control the quill extension. This makes it possible to pre-set to drill any number of holes to exactly the same depth.

When drilling through, use scrap wood between work and drill-press table. Hold the work firmly to keep from twisting, or use clamps. Here 1½" spade bit is being used.

With special high-speed accessories, you can use the drill press as a pretty efficient shaper. Cuts here are samples of the kind of thing you can do with the shapers.

Some tools, like dovetail bit shown, are held in special holders replacing regular chuck. Necessary when the cutter develops side thrust. Another example: router bits.

To drill holes through round stock or tubing, make a V-block jig to hold the work. Line up drilling tool with bottom of the V so pieces will be drilled diametrically.

Many sizes of drum sanders can be used on a drill press. To sand wood edges square, make a table that drum can pass through. Auxiliary table is clamped to drill table.

It's handy to have a drill-press table like the one shown that can pivot. Then you can establish a plane that is parallel to the to the spindle center line of drill press.

A portable drill that has both speed control and a reversing switch is something to aim for. You can use it at low speeds to drill masonry and for turning screws.

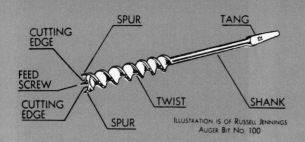

CUTTING EDGE · SPUR · TANG · FEED SCREW · CUTTING EDGE · SPUR · TWIST · SHANK

ILLUSTRATION IS OF RUSSELL JENNINGS AUGER BIT No. 100

Nomenclature of an auger bit. This is used in hand brace only, never in a power drill. Auger bits sized by 16ths of an inch, measuring the diameter. Lengths vary from 7-10". Dowel bits are short auger bits.

Spade bits are great for drilling of holes in wood regardless of whether you are working with a drill press or a portable drill. Work best about 1500 rpms.

Be aware of difference in bit points. Spade bit can be used under power, not so the screw-tipped auger; this bit is designed for use in hand braces only.

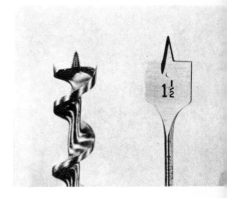

units would have been required. Some models not only have speed control but include a reversing switch. This means, for example, that you can even drill and remove screws with the same tool you use to drill a half-inch hole in masonry or a 1/8" hole in wood.

DRILL SIZES

Sizes are figured in terms of chuck capacity. A ¼" drill will grip shanks up to ¼" in diameter. A ½" drill will grip shanks up to ½" in diameter. This doesn't mean that a ¼" drill can't bore holes larger than ¼". Spade bits, for example, that you can buy for holes as large as 1½" have ¼" shanks so they can be gripped in a ¼" drill. This holds true for other tools like hole saws, sanding discs, etc. Still, it stands to reason that the smaller the chuck size the less power and bulk the tool will have. You don't want to go overboard in either direction so I feel a 3/8" size with speed control makes an awful lot of sense, at least for the very first one you buy. If you buy a good one

(don't try to save a couple of bucks here) you may never buy another.

Like the stationary drill press, there are many accessories available for a portable drill. These range from polishing wheels to saw so that you can actually build a small scale power shop around a portable drill. If you are inclined this way, check the accessories available for the unit you are interested in before you buy.

HAND DRILLS

If you go the hand tool way—start with at any rate—you want a ratchet brace and a set of screwtipped bits. The screw helps pull the bit into the wood as you turn the brace. If you are doing furniture work you should be able to form holes up to ½" in diameter. Bits above this size can be purchased as they are needed. In a sense, the brace is a "heavy-duty" tool that can be used efficiently on large holes. For lighter work, a hand drill is better. The chuck on this unit will hold small bits and twist drills.

27

Disc sander is either an individual tool powered by its own motor or an accessory disc mounted on another machine. Radial arm saw, lathe and even table saw will take a disc.

TOOLS FOR SANDING

How to select the right tool and the correct abrasive for the job

In the stationary tool catagory you have the "disc sander" and the "belt sander". You can buy these as individual items or as one machine that combines both. The latter makes sense since you can use the same stand and the same motor for both.

THE DISC SANDER

The disc sander is not much more than a flat steel or aluminum plate to which you attach a circular sheet of sandpaper. Since the sanding action is rotary, it is not the best tool for surface sanding since it will leave cross-grain marks. However, it is excellent for smoothing end grain, touching up miter cuts, finishing outside curves, smoothing or actually forming chamfers and doing other jobs of that nature.

THE BELT SANDER

The belt sander runs an endless abrasive belt and you can apply work so that sanding is done in line with the grain.

Both of them are especially good when you have to smooth down or shape small pieces. You can use them to smooth down the edges of a large piece of plywood but not the surface. If you wanted to smooth down the surface of a piece of lumber 4″ × 12″ you could do it very efficiently on a belt sander but if the piece was 12″ × 36″ it would be best to seek other means of doing it.

I'm not trying to belittle the functions of these tools in a shop. There's little doubt that once you have them they will make many tasks easier to do but, in a general sense and in terms of the way to go when you are acquiring your first power-sanding equipment, we have to give more attention to the portable ones.

THE PAD SANDER

In the portable area we have the "pad sander" which is thought of as being a finishing sander, and the "belt sander" which is more-or-less a workhorse. If you do woodworking, having one of each of these should be considered

If you have no choice but to start with one, give a lot of consideration to the pad sander. It's a very flexible tool in terms of application and most of the materials you work with are fairly smooth to begin with so that a lot of heavy sanding isn't usually necessary.

It's really fairly rough to try and improvise a belt sander — even this one above, mounted on a Shopsmith is pretty much an individual tool that can stand by itself.

Many companies, like Stanley, make several models of pad sanders. One here has built-in dust collection system, larger pad, faster orbit. Cost is twice that of the other.

Pad sander fitted with coarse abrasive can be worked crossgrain for faster material removal. Final sanding must be done with the grain with fine abrasive.

Pad sander is all-around finishing tool but it shines on jobs where it would be difficult to control a belt-sander. The Rockwell "Speed Bloc" is the best around.

Pad sander can get into those places where a belt sander would fear to tread. Good action from a pad sander does not require any great amount of pressure on the work.

Pad sander is good sharpening tool. To accomplish what amounts to fine honing job, place piece of ⅛" hardbaord between pad and paper— very fine, wet-or-dry paper.

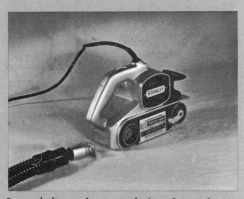

Some belt sanders are designed to take an accessory attachment and hose so you can use a household vacuum to collect the dust. Empty the bag often to prevent clogging up.

Others are made so you can attach a dust-collection unit directly to the machine. The attachment, however, is an accessory you'll have to pay a little more money for.

Check belt sander you're interested in for projections on open side of the belt. This makes it possible to work directly up to a vertical surface. Good for re-doing floors.

Be careful at the end of a stroke with a belt sander to guard against tipping the machine. Come off on horizontal plane so you are able to avoid rounding off edges.

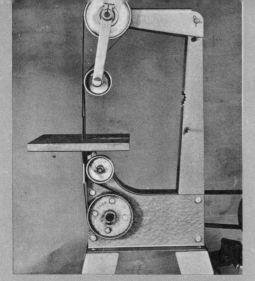

"Belt grinder" runs a narrow, endless abrasive belt so you can slip belt through opening in work before mounting on machine. Thus you smooth edges on internal cutouts.

There are many other tools in a workshop that can also be used for various sanding jobs. For example, a drum sander on a radial arm. Both practical and economical.

WHICH TO USE?

There *is* functional overlap but there are also some decisive factors that pertain more to the physical characteristics of the job than to the results required. If you have a flat sheet of lumber or plywood in average from-the-yard condition, either tool could be used to do the final smoothing. The belt sander would get there faster. In a down-to-the-wire test, the pad sander would produce the smoothest finish because it will drive finer abrasives than on a belt sander.

If you had just glued up a slab from individual boards that were less than exact in thickness and where some slight warp could not be eliminated before the clamps were applied, a belt sander would be *the* tool to use to bring the job down to uniform thickness. You could continue and finish the job with the belt sander, but if you also owned a pad sander you would take up with it from that point.

If the job involved assemblies, inset panels, sculptured areas, rounded corners and edges, smoothing without heavy material removal, then the pad sander would be more applicable and would do a fine job in some areas where a belt machine might do more harm than good.

When you think in terms of smoothing a floor or a deck, even a concrete or flagstone slab, then you want a belt sander. For the final touch on the parts for a coffee table, a paneled door or a project made from plywoods with a fine surface veneer, think in terms of the pad sander.

THE CORRECT ABRASIVE

No sander will do its job unless you use the correct abrasive. The general rule is to work down through progressively finer grits until you achieve the finish you want. But that doesn't mean going from grade to grade from the coarsest to the finest. Start from the finest grit possible in relation to the condition of the work and then skip two or three grades at a time.

Use an "open-grain" paper when you are removing finishes or working on soft or resinous woods. Use a "close-grain" paper on all normal jobs.

Aluminum oxide is an excellent abrasive for power-tool use. It costs more than garnet (which is quite popular) but in terms of surface sanded per penny spent the aluminum oxide will turn out to be more economical in the long run.

TOOLS FOR PLANING

Your hand plane is fine, but powered tools will cut work in half

The most popular tool in this area and one quite common in home workshops is the "jointer". It does its job with a rotary cutterhead mounted between an infeed and an outfeed table. The forward table is adjustable for depth-of-cut; the other is usually fixed. When you move the work forward over the cutterhead, it removes the amount of material you have adjusted for. A vertical fence, almost as long as the combined lengths of the horizontal tables, provides support for the work and can be adjusted so you can hold work at an angle while passing it over the blades. Thus you can do chamfering and beveling.

Quite often you will see a jointer set up in combination with a table saw—sometimes both operating off the same motor. This is a nice setup because it lets you smooth down edges that you cut on the saw. Of course the same benefit is available when the tools are mounted separately.

A 4" jointer is about average home-workshop size. Not because it's ideal but because it falls within the homeworkshop price range. A 6" jointer is huskier and has more capacity and longer tables but both sizes work the same way. Depth-of-cut is an important factor—not for planing—but because the jointer is often used for rabbeting. Assuming that all other features between several models are the same, choose the one with the greatest depth-of-cut.

SURFACE CUTS

You can make surface cuts, even thin down stock, but don't confuse a jointer with a thickness planer. The latter is a big, expensive machine seldom found in a homeworkshop but it will plane rough

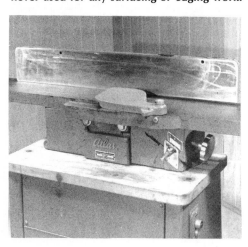

The heavy-duty jointer shown has 6" blades and a ½" depth-of-cut. Big depth-of-cut is fine for rabbeting and similar cuts but is never used for any surfacing or edging work.

Rabbeting is accomplished by adjusting the fence for the width of cut, infeed table for its depth. Notice that blade guard is removed for these jobs, so watch fingers.

When you rest the work on the edge of the outfeed table before making a pass, you form a taper. Good for making furniture legs. Stop blocks clamped to table control lengths.

stock to precise thicknesses with the assurance that opposite surfaces will be parallel to each other. Doing this on a jointer is more a question of skill in using the machine—and—the width of stock you can handle is limited mostly to the length of the blades. A thickness planer will handle stock at least 12″ wide and as much as 6″ thick.

Once you own a jointer you will find yourself using it for many jobs other than planing edges. Rabbeting for joints, taper cuts for furniture legs, tenons, bevels, chamfers, are among the standard woodworking techniques easily accomplished on a jointer.

THE UNIPLANE

A newcomer in this field is the Rockwell-Delta *Uniplane*. It much resembles

The Uniplane resembles a jointer, but it really is a different type tool. Infeed and outfeed tables are vertical. Special cutters revolve around the fixed center disc.

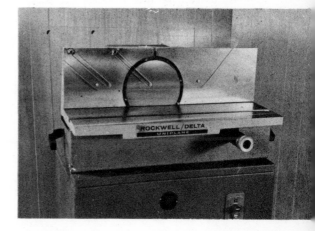

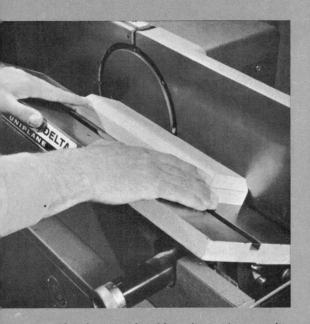

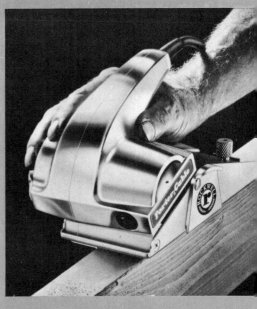

The horizontal table adjusts for angular cuts such as this chamfering job. Maximum depth of cut is ⅛". Going beyond this is just a matter of making repeat cut passes.

A power plane is just that, a special tool made to take the hard work out of jobs you might otherwise accomplish with your hand plane. It can be used on surfaces or edges.

a jointer but it does its cutting with a rotary planer type head that operates in a vertical position. The fine cutters in the head are so set that the machine actually makes two cuts. A stock removal cut when you start the feed; an extremely fine finish cut as the wood leaves the cutting area. This feature plus a slow feed gives extremely fine finishes on edges and surfaces and even stock ends. You can actually smooth down the end of a 6×6. Also—very unlike a jointer—you can work on very small pieces of stock. It's quite safely used to plane down material as small as ¼" square. The catch is, it's pretty expensive. The only version available right now is close to $400.

PORTABLE POWER-PLANING

For portable power-planing, you can go two ways. A typical special one made for the purpose is shown in the photo. It can take a lot of the drudgery out of

jobs requiring a lot of planing (such as smoothing down rough stock or reducing stock thickness) and it can be used to do touch up work or finishing work on edges. Thus it can be used, for example, to fit doors or to dress down sticky ones. With an adjustable guide mounted, the tool can cut or dress bevels or chamfers. Anything you can do with a hand-plane you can do with this tool but with a lot less effort and considerably more speed. There are larger versions available than the one shown but this one-hand controlable model is a good one to check out for the home workshop.

THE PORTABLE ROUTER

The second way to go if you own a portable router, is to buy an accessory for it that will do planing jobs. The restriction here is that you'll be able to work on edges only, but that can cover a lot of jobs in the shop and in the house,

You can buy an attachment for a portable router, but planing will be limited to the edges of work. This, however, will prove be the bulk of such work you will be doing.

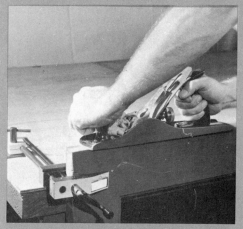

Hand planing calls for holding work firmly and making flat, parallel passes in direction of the grain. With a light cut and a sharp blade, waste will curl off in strips.

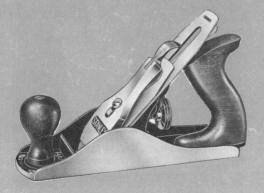

There are a variety of hand planes, but a 12″ to 14″ jack plane (plus a small block plane) will set you up nicely for most of the usual home handyman workshop projects.

especially when you consider that most of us work with materials that are already surfaced anyway. The accessory includes an adjustable fence so the combination can be used to do beveling and chamfering as well as full edge-cutting.

HAND PLANES

Hand planes do the same jobs described for the above tools but you use muscle-power instead of electricity. And of course you have to spend considerably less money to acquire some. A 12″ to 14″ jack plane is a good place to start. With this tool you can smooth or dress down edges and even do surfacing. A lot of material can be removed quickly when you work the tool at an angle across the grain. Thus, smoothing down of rough stock—even thinning—is not out of line. Best way to proceed is to gradually reduce the crossgrain angle until you are doing the fine passes parallel to the grain

and cutting in the grain direction. You'll know when you are not cutting with the grain because you will be lifting splinters and chunks instead of smooth shavings. On most of the wood you work with, grain direction will not be difficult to determine beforehand. The grain pattern shows up as arrow-like formations. Just work in the direction the "arrows" point. On edges, you do the opposite.

THE BLOCK PLANE

A small block plane is a good addition. This is a one-hand tool that you can use on stock ends by making passes from each side toward the center. Working full-length will break splinters from the wood at the end of the stroke. Cuts like this should always be done with minimum blade projection. Use a block plane too, for chamfering and for breaking corners.

Spindle turning on a lathe. The tool rest supports the chisel as you apply it to work. Good lathe work depends on the skill with which you apply tools. Practice is the key.

SHAPING & TURNING

Honest appraisal of the value of owning a lathe, shaper or router

In the stationary tool catagory you have the *lathe* and the *shaper*. The difference between the two is this—on a lathe you do circumferential forming while the shaper is designed primarily so you can work on stock edges. You can turn a table leg on the lathe or make a lamp base. On a shaper you can form a cabinet-door lip or a drop-leaf table edge. These are typical jobs exclusive with each of the tools.

THE LATHE

On the lathe, columnar projects are formed by mounting the work between centers. This is called "spindle" turning. Flat projects, like trays and bowls, are mounted on a flat plate   and this is called "face plate" turning. Capacity is determined by the distance between centers and twice the distance from the spindle center to the lathe bed. A 12″ lathe, for example, indicates that you can face-plate form a bowl 12″ in diameter. If the length is 36″, that's the limit of the work you can mount between the centers. For work beyond this you would form the piece in sections and assemble with a dowel joint.

A "FUN" TOOL

Generally speaking, the lathe is not a very essential tool these days, especially when you think in terms of home mainte-

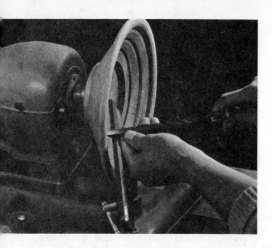

This is face-plate turning. The work blank shown was built up by laminating rings of stock to reduce waste to be removed. Lathe round-nosed chisel is used to rough work.

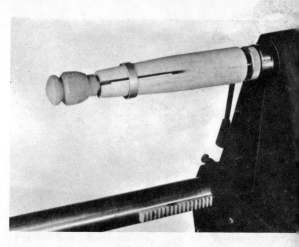

After you have gained some experience on the lathe you can create setups like this. A home-made wooden chuck mounted on a Shop-Smith spindle in lathe position holds work.

A small example of the kind of thing you can turn out on a lathe. It's a lot of fun to see a block of wood gradually transform into such attractive household accessories.

nance and general carpentry. Modern furniture styles have departed from turned legs and even should your design incorporate some, chances are you can buy ready-made sets that will be suitable. Actually, the decision about getting one will be based mostly on whether you will enjoy the work. There is no doubt that the lathe is a great "fun" tool and if you are looking for something for pure enjoyment (and there's nothing wrong with that!) the lathe would not be a bad choice. It's one tool on which you can accomplish complete projects—such as, candle sticks, lamp bases, bowls, trays, turned boxes with fitted covers, and so on. By making turning blanks by gluing pieces of contrasting woods you can create very intriguing projects.

THE SHAPER

The shaper operates by turning cutters that are mounted on a vertical spindle. This is another tool that isn't seen too often in a homeworkshop but, in this case, because many of the jobs it does can be accomplished with accessory equipment on a drill press and even a table saw.

Infeed fence is set for depth of cut, the outfeed supports after cut. Fences are kept in line when only part of edge is removed; when all is removed, outfeed compensates.

Blank cutters are available to remove the entire edge of the stock (like a jointing cut) or by height adjustment, partial cut, to form rabbets. Make tongue by flipping.

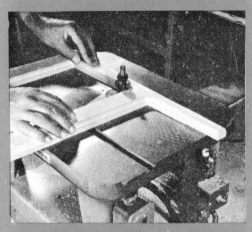

When work size permits, do shaping after partial assembly for smooth flow of lines in joint areas. Depth-of-cut is controlled by collars mounted on spindle with cutters.

Two late model routers made especially for home workshop use. Each one sells for under $40.00. Don't let the size fool you. They are both capable of real professional work.

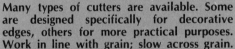

Many types of cutters are available. Some are designed specifically for decorative edges, others for more practical purposes. Work in line with grain; slow across grain.

You can form dadoes or grooves with blank cutter, but use guide at first; a strip of wood clamped in place so you can move the router along it as you are making the cut.

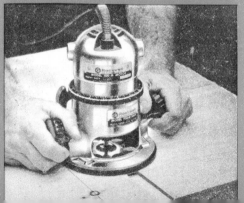

It's a high speed machine and should be well powered so it can produce the smooth cuts it was designed for. In the area of speed and capacity it has it over an accessory-equipped drill press. For one thing its spindle projects from *under* the table so you have nothing to hinder you from above as you do on the drill press. For another, most drill-presses will not stand up under the high speeds built into the shaper; and it's high speed that makes smooth shaper cuts possible.

Don't regard the shaper as a tool designed to produce decorative edges only. You would be selling it short. With the right cutters it will form cabinet-door lips, do rabbeting, form tongue-and-groove joints, drop-leaf table edges, special glue joints, form the shapes for window sash including the cope, do panel raising and many, many other standard woodworking chores.

THE ROUTER

In the portable area you have the router and it's a dandy tool to own especially today when some good models are available at reasonable prices. In essence, it's a shaper that you can hold in your hand to use on any job, anywhere, in or out of the shop; 20,000 to 30,000 rpms are not uncommon in a router and this produces extremely smooth machined surfaces that require no further attention. The router, like the shaper, does much more than form decorative edges. It will, with cutters only or with added attachments, do the following jobs: Mortising for hinges, dovetailing, trim laminates on counter tops, make the cuts for inlay work, incise or raise figures, do piercing, etc. It will also do the same kind of joint-forming operations we outlined for the shaper.

A fascinating feature of the router is that you can apply it to a completely assembled project to add a professional touch. Most good cabinet shops consider the router an extremely important tool— and we consider this a pretty good recommendation.

Router can also be used to follow curved edge. Cuts made in plywood will never be smooth as those made in solid stock, but a slow feed helps through more cutter rpms.

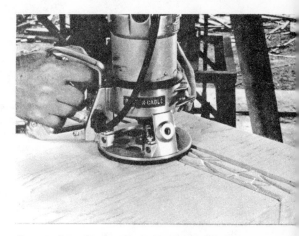

Decorative, freehand surface cuts are easy to do. This is the way to raise letters for name plates or house signs. Be careful on crossgrain since grain will tend to guide.

For most routers you can buy an accessory table. This makes it possible to set up and use your portable router as if it were a small size shaper. Handy for many jobs.

Buy a good hammer, one that feels right in your hand. It will last a lifetime with a little loving care. A 16 oz. model is good for general use. For jobs like this, a 20 oz.

HAMMERS & SCREWDRIVERS

How to select and use these tools is information you should know

A hammer is a very personal thing; you build up a kind of buddy relationship with it. That's why choosing one, assuming that you are looking at an assortment of *quality* products, is much a matter of how it feels in your hand. Grip firmly about one inch from the end of the handle. Test its balance and feel by driving an imaginary nail. Small nails can be driven through a wrist action only with your grip closer to the hammer's head; larger nails bring the elbow and shoulder into play with the grip very close to the end of the handle. The hammer for you should feel good no matter how you use it.

Also, remember these facts when you buy. A hammer may have a "bell face" or a "plain face". The striking surface of the *bell* is slightly convex; this so you can drive nails flush without marring adjacent surfaces. This is a good feature but it makes the hammer a bit more difficult to use because the slightly rounded face can slip off a nail head more easily than a *plain* face. The plain face is flat and most beginners find it easier to use.

The hammer can have a curved claw or a straight claw and there are some that fall somewhere in between. The curved claw is best for drawing nails. The straight claw can be used that way too but it's designed primarily for separating pieces of wood. It's easier to insert between boards and it provides a good lever action. It's not a good idea, however, to use it this way on, for example, heavy framing members. A wrecking bar is the tool for that.

HAMMER SIZE

The hammer "size" is figured by the weight of the hammer's head—they run from 7 oz. to 28 oz. For general cabinet work and furniture making, choose one in the 16 oz. catagory. If you are going to do heavier work such as house framing—maybe even fence building—you can go to 20 oz. Professional house-framers use a "framing" hammer. This is heavier, has a longer handle and a corrugated face. The extra weight and leverage makes it easier to drive heavy nails—the rough face minimizes the possibility of slipping off the nail head when delivering heavy blows. If the corrugations should leave an occasional scar—it doesn't much matter on rough framing.

In use, train yourself to strike with the center of the face. Most hammers are specially hardened in that area to take the abuse. Don't hammer with the side of the head. Avoid striking materials that are harder than the steel in the hammer. Be very careful when driving case-hardened nails or use a light sledge. When pulling nails, especially large ones, place a block of wood under the hammer head. This acts as a fulcrum for extra leverage.

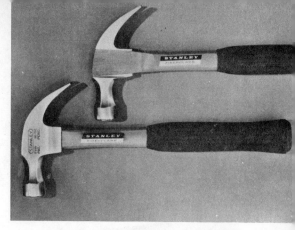

This modern version features a fiberglass handle and a "taper-lock" assembly that is further strengthened with special adhesive. It's a 16 oz. claw model, best for pulling.

Don't work with a hammer that has a fractured handle or a loose head and don't try to use tape as a temporary cure. Replace the handle and/or drive new wedges to secure the head. This can't happen to some of the modern hammers since they are practically one-piece affairs.

ABOUT SCREWDRIVERS

The most important thing to remember about screwdrivers is to buy them in sets; this because one or two will not do for all size screws. The tip of the blade should fit the slot in the screw-head. If the tip is too small it can bend or break. If it is too large it won't fit at all or you may get a partial grip with the possibility of damaging the screw and the work.

Also, since the tip must be larger along with the increase in size of the screw head, the driver itself is larger so you can apply more torque.

Special types include insulated models for electrical work, offsets for getting into tight places, stubbies for close quarters, and some with screw-head grippers mounted on the shank. These are great for getting a screw started when you can't do it with your fingers.

Square shank screwdrivers fall in the heavy-duty catagory. With them, you can grip the shank with a wrench when you need an assist to do the turning.

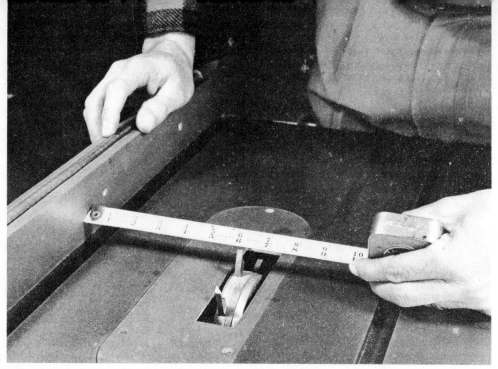

A good flex tape is the first measuring tool to buy. It should be at least eight feet long, easy to read. You can buy tapes up to 100-ft. long for larger outside projects.

MEASURING TOOLS

Levels, squares and dividers are almost as important as your rule

The advice, "Measure twice—cut once," is an old but very wise rule to follow. It can save a lot of material and avoid a lot of frustration. The only way to use measuring tools incorrectly is to use them carelessly.

The items you should acquire quickly are those shown in the illustrations. The following are others that you can add if and when you need them.

The steel FRAMING SQUARE (often called a carpenter's square) has two arms that form a right angle. The longer arm is called the "blade", the shorter one the "tongue". A regular one, in addition to normal graduations, carries markings that are useful in rafter and framing work. It makes a good straight-edge; the blade is 24″ long and it's good to use on jobs where the smaller adjustable square is not adequate.

A WING DIVIDER or compass. This can have two steel points or one steel point and provision for holding a pencil. In addition to drawing circles you can use it to step off measurements, to transfer a measurement from one point to another, to mark off equal spaces, draw lines to match irregular surfaces and do decorative work like laying out scallops.

A V-HEAD for an adjustable square is used to locate centers on round stock. You would use it, for example, before mounting a spindle for lathe turning. It can also be used to locate the center on square stock.

A small LINE LEVEL. This hooks to any stretched string so you can determine levelness or grade. Good for laying out patios, foundations, brickwork, even for trimming hedges.

Stanley design has a hook so you can keep the tape handy on your belt. There is also a thumb-operated latch on this model so you can keep tape locked in extended position.

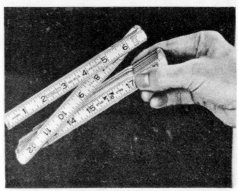

A zig-zag rule can be extended to six ft. and then folded down to six inches. These are available in wood or metal. Some have special graduations such as brick spacing.

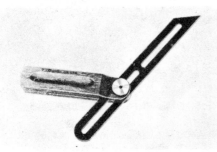

Adjustable sliding T-bevel is used to lay out angles or to check existing ones. You would use T-bevel to measure angle of a pitched ceiling when installing molding.

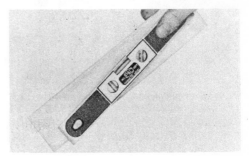

A level is important for doing jobs that can't be checked with a square. You establish alignment on fence posts and house-framing posts by using level on two faces.

New all-in-one tool draws lines, circles, gauges nails, screws and dowels, can be a level or square; even ruler is removable.

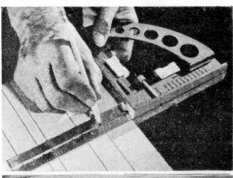

This handy tool is both a chalk line and a plumb bob. The chalk line is also handy indoors for the snapping of wall-length lines.

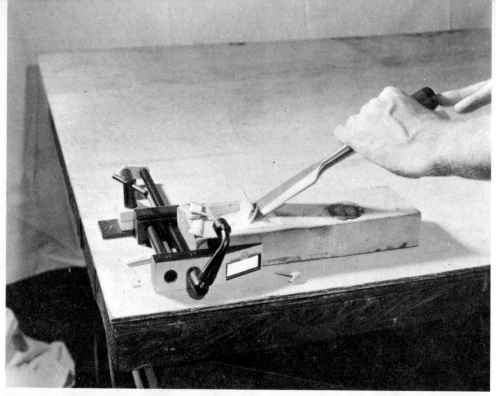

A set of woodworking chisels are a must in the hand-tool shop. Use them for hinge mortising, joint forming and the like. The set should run from approximately ¼" up to 1".

MISCELLANEOUS TOOLS

Here are the extras you may need to round out your standard tools

In addition to the tools already discussed, there is a long list of products that can help you do better or faster woodworking—or serve as helpers when you do routine home maintenance jobs. Individually, they are not too costly; as a buy-at-one-time kit they represent a considerable investment. Best bet is to check through this section so you'll know what's available and then buy them. In addition to those illustrated, consider the following.

A 14" pipe wrench—not a woodworking tool but needed for an occasional indoor or outdoor plumbing job.

A set of combination wrenches—one end open, the other end boxed. This will let you handle nuts that are square or hex shaped. You'll need them for lag screws as well as nut-and-bolt assemblies.

Diagonal cutters—use them for cutting wire, not on nails or screws.

An adjustable (10") wrench—use it in combination with the wrench set or on jobs where the largest piece in the wrench set won't fit.

One of the long wrecking bars—this will have a pry at one end for dismantling structures or prying boards apart; a nail puller at the other end so you won't be tempted to use your claw hammer on spikes and similar fasteners.

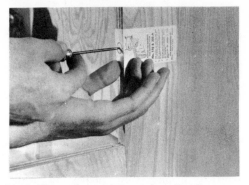

Mark dimension points and scribe lines on soft metals with awl. It's often the only pilot-hole needed to drive screws in soft woods. Use it to mark screw locations too.

"Surform," a type of wood rasp, comes in many shapes and sizes. Good for shaping wood. It cuts on the forward stroke and should be used in line with the woodgrain.

Metal cutting files are good for smoothing edges and for sharpening your other tools. They should not, however, be used on wood since most soft materials will clog them.

A saw-knife tool can be used like a key-hole saw on both soft materials and wood. With the correct blade, it is also good for doing light-duty cuts on most metals.

A stapling gun can't replace a hammer but it can make many jobs easier and faster to do. Choose the correct length staple for the material you are planning to secure.

Photo shows a countersink that's good for doing a neat job when you are using flat-head wood screws. Those for wood have a different angle than those for metal working.

Putty knife for re-glazing windows. Also available in 3″ size — which you would use should you become involved in taping plasterboard. Don't use it in place of scraper.

Propane torch, as appears in photograph, comes with a set of tips that can be used for many jobs — from sweating copper pipe joints to helping you remove old paint.

Sheet metal snips come in three ways according to the particular cut you desire. There's the "combination pattern," the "regular," and "duckbill" to choose from.

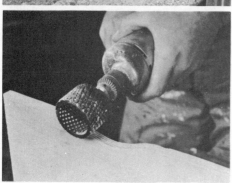

Rotary types available for use in a drill press or with portable drill. Clean cutting teeth often, especially when using on softwoods. Don't work them at high speeds.

Cold chisels may be low on list of tools for woodworking shop, but if you ever have to remove head or nut from rusted bolt that can't be turned, this is item to do it with.

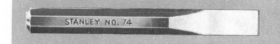

STANLEY NO. 74

Adjustable pliers can be set to suit many different size jobs. Long handles provide a lot of leverage. Regular slip-joint pliers can be used nicely for light-duty jobs.

Lever wrench pliers (or "vise-grips) are adjustable for different size jobs and will lock in place when the handles are closed. Almost like a vise you hold in your hand.

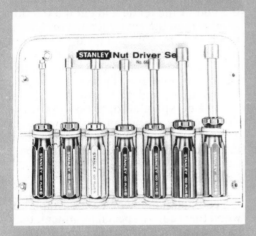

Buying one won't do you much good by itself, so consider sets when you are shopping. There is also a type made that can easily be adjusted for various size nuts.

Tube cutter may seem far out for woodwork shop but if you ever want to do something with do-it-yourself aluminum tubing you will enjoy having this tool to work with.

As shown, a good easy-to-handle knife can be useful in many situations. This one is made so you can replace the cutter. In fact handle holds three or four extra blades.

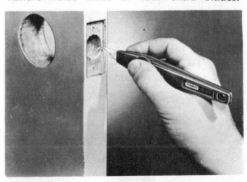

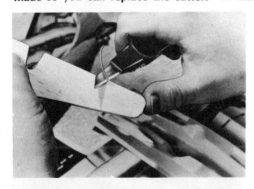

CLAMPS AND ADHESIVES

There are many kinds, so you must choose correctly to suit the job

These days you can trigger glue from an electric gun, spray it from a can, squeeze it from a tube or press it from a plastic bottle. There are still adhesives you apply by brush and some that come in powder form for you to mix with water. It stands to reason that application procedures will vary somewhat and that points up the importance of following the instructions on the container.

If the instructions tell you to coat *both* surfaces, to wait for the glue to get tacky before you join the parts, to keep under clamp pressure for a specific period of time, to avoid use below certain temperatures—it's because that's the best way to get the most from the product.

OTHER CONSIDERATIONS

Check the list at the end of this section and choose the adhesive that is best for the job you are doing.

Parts shaped for joining should fit together in fine fashion *without* glue. Glue does not compensate for a poor fit. This holds for simple as well as complex joints.

Since glue has body you must provide room for it in such joints as the mortise-tenon and the dowel. Always chamfer the end of the tenon and cut it a fraction shorter than the cavity it must fit. Use as much glue as you need for the job. Even with the precautions mentioned, excess glue may not be able to escape and this can cause problems. Dowel ends should be chamfered and the length of the dowel should be spiraled or grooved for the same reason. You can buy dowel pegs all prepared for the job but you must drill the holes to suit the size dowel you buy.

SPECIAL TREATMENT

Teak and lemonwood are examples of wood that require special treatment; they have a surface film that can interfere with the action of some types of glue. But *casein* glue is available to solve that problem.

Contact cement should be used for

Thermogrip gun shown is one of the latest products to help make gluing easier. It works electrically to melt sticks of glue. Bond sets in 20 seconds and is waterproof.

Spray adhesives are available but mostly for non-wood work. With these, follow instructions carefullly. Work in an open area so you can prevent damage from over-spray.

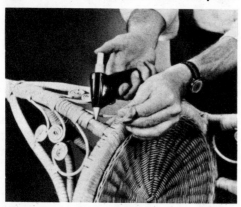

The gun can be used for many jobs around the house — including caulking with special sticks instead of the glue. Larger version of above gun is now available. Handy to use.

Many types of cement come in small tubes — to mend anything from chinaware to plastic parts. Some of these are waterproof, some not. Note use of handy spring-type clamp.

Smaller version of bar clamp has deeper jaws. Clamp hold-down is accessory that hooks to bar and keeps broad slabs from bulking under any pressure from the clamp.

Bar clamps are excellent for furniture builders. They come as units and in specific lengths. This is bad place to save money — try to buy only really good ones.

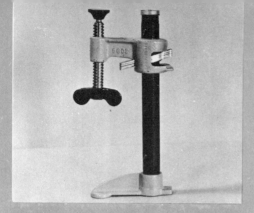

Pipe clamp fixtures are more economical for homecraftsmen. You buy the screw assembly and heel; supply the pipe yourself. Note—they push out as well as push in.

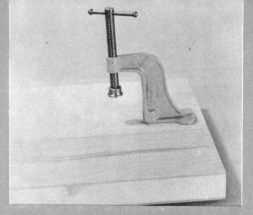

Hold-down clamp can be used on a work bench or a drill-press table. It is removable so one clamp can serve both purposes. Clamping work will insure both accuracy and safety.

With a few sets of pipe clamp fixtures and assortment of varied length pipe, you can do variety of jobs—even like this. Black pipe is better to use than galvanized pipe.

Web clamps or band clamps are great for jobs like this where it would really be difficult to use conventional clamps. Both are also good for clamping round objects.

lamination jobs, not for securing conventional joints. When using it on porous materials, let a first coat dry, then apply a second one. Be sure to wait the correct drying time before bonding the parts.

For outdoor work—any project that will be exposed to moisture—you use a waterproof glue.

Use only as much glue as you need for the job. Applying more is just waste and extra work since you must clean the excess away. Some squeeze-out is unavoidable and this should be removed immediately. Dry glue is difficult to remove and, if allowed to remain, will show up under stain.

When doing edge-to-edge joints be sure the edges are flat and square to begin with. As we said, don't depend on the glue to hide gaps. On hard, very dense woods, the glue-joint will be stronger if you work the mating edges crossgrain with fine sandpaper.

Heat will help some glues set faster— and, if this is so with the product you are using, the label will tell you so. In these

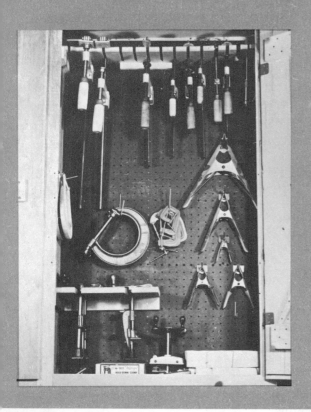

Build up assortment of clamps, but buy good ones so they last. The C-clamp is a general-purpose tool. When using on wood, remember to place scrap-paper between the work and the clamp.

instances and especially on small projects, a heat lamp can be used very effectively. Most times, glue will set faster and better if you work in a warm room. This doesn't mean you can't work unless the temperature is 70°; there are upper and lower extremes.

THE MOST POPULAR ADHESIVES

LIQUID RESIN—is common and comes ready to use by squeezing from a plastic bottle. Sets pretty fast under clamps. Can be used on materials like leather and cork in addition to wood. It will resist moisture but it is not waterproof.

CONTACT CEMENT—is usually applied by brush or with a spreader. It does what it says—bonds on contact—so you have to be careful when joining parts to get them right the first time. You can use wrapping paper between large sheets (after the cement is dry) and then pull the paper away when you are sure that the pieces are aligned. Contact cement is ideal for attaching laminates to counter tops, joining panels, and for similar jobs. It is not used on ordinary joint work.

POWDERED RESIN—it's mixed with water but only in quantities needed for the job on hand. It requires clamping; it's strong, and it's waterproof. The glue to use on that outdoor lounge or table.

RESORCINOL—at the top of the list of waterproof glues. It requires mixing and secure clamping and sets better when worked at temperatures above 70°. It is very strong when used correctly and recommended for projects to be used outdoors, on boats and so on.

CASEIN (POWDERED)—requires mixing and can be used for general woodworking but it's *the* glue to use on oily woods. It is water *resistant*.

ANIMAL AND FISH GLUES—seldom found in the homeworkshop these days even though they are strong and good to use on furniture projects. Should be used warm. Not for projects exposed to moisture.

NAILS
AND
SCREWS

Special nails do special chores, but screws look and hold better

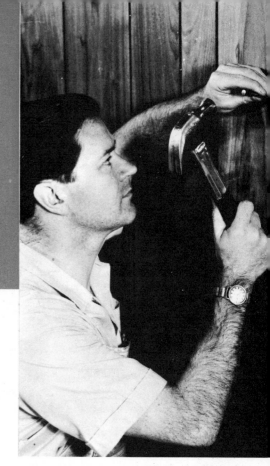

Nail set is used to drive finishing nails lower than the work surface on paneling, as shown, above. Use the right nail set, however, to keep the hole as small as possible.

Of all the nails available there are two types that you will be using for general purpose work. These are the *common nail* and the *finishing nail*. A big difference between the two is in appearance on the project. The common nail has a broad head and is driven flush with the surface of the wood. The finishing nail has a smaller head; really not much bigger than the shank, and it is usually set below the surface of the wood so you can hide it with a putty. So—in general—when choosing the two, decide whether it matters if the nail shows. The common nail is stronger when you think in terms of pulling away the nailed piece. Its head has a bigger grip than the one on a finishing nail.

SPECIAL NAILS

In addition to these there are many special types—items designed to do a better job on a specific material or application. To become familiar with these is quite a task and not really necessary since, more often than not, the name of the nail describes its use. Thus any special job you do will identify the fastener you need.

For example—there are sheetrock nails, siding nails, corrugated roofing nails, asphalt-roofing nails, gutter spikes, flooring brads, metal-lath nails, scaffold nails and so on. When you are puzzled about the right nail to choose simply give it the name of the job you are doing and/or the material you are working with and the dealer will know exactly what you want. The thickness of the materials will decide nail-size. A general rule states that the nail should be three times as long as the thickness of the material you are nailing through. Thus if you are securing ½″ stock the nail should be 1½″ long. Sometimes this can't apply—suppose you nail two pieces of ½″ stock surface-to-surface? The "general rule" nail would be too long. In such a case you would use nails just long enough to do the job without protruding—or use screws.

ACTUAL SIZE OF A 60d 6" COMMON NAIL

OTHER SIZES	
50d	5½"
40d	5"
30d	4½"
20d	3¾"
16d	3½"
12d	3¼"
10d	3"
8d	2½"
6d	2"
4d	1½"
3d	1¼"
2d	1"

APP. NUMBER OF NAILS PER POUND

d SIZE	COMMON	FINISHING	CASING
60	10		
50	13		
40	17		
30	20		
20	30		
16	45		75
12	60		
10	65	125	94
8	100	196	149
6	165	309	245
4	290	600	485
3	540	875	
2	845		

These nails sold in bulk, although you can buy some in packages.
Brads are sold by length and gauge—usually in packages.

Special threaded nails are very good when you want to nail down underlayment and be sure it is secure. They hold better than smooth-shank nails, guard against popping.

COATED NAILS

Nails are also coated or tempered or made of special materials. Case-hardened nails are used to attach materials to masonry—you drive them as if you were going through wood. Coated nails (dipped in cement or maybe etched or barbed) are used where there is a possibility the nail might pop. Typical applications are securing under-layment or sheetrock.

RUST-PROOF NAILS

Copper, aluminum, cad-plated or galvanized nails are available for places where rust and discoloration can occur. Aluminum nails, for example, do a good

53

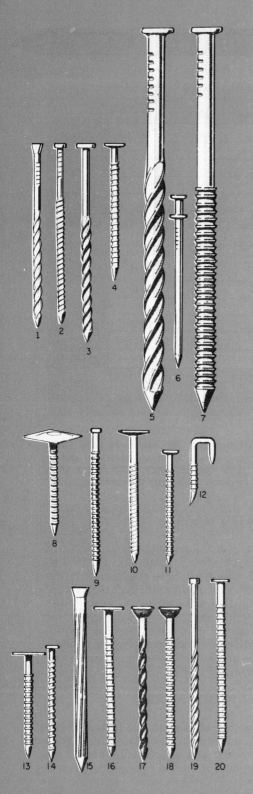

1
2
3
4
5
6
7
8
9
10
11
12
13 14 15 16 17 18 19 20

NAIL	EXAMPLE USE	NAIL-MATERIAL
1-2	Siding	Depends on Siding
3-4	Sheathing	Steel
5	Framing	Steel
6	Temporary Work	Steel
7	Rafter Anchor	Steel
8	Roll Roofing	Varies
9	Paneling & Trim	Varies
10	Drywall	Steel
11	Flashing, Gutters	Copper
12	Wire Fencing	Steel
13	Asphalt Shingle	Varies
14	Asbestos Shingle	Varies
15	Anchor to Masonry	Steel
16	Wood Shingle	Varies
17-18	Sheet Metal	Varies
19	Flooring	Steel
20	Subfloor Underlay	Steel

job on wood siding since they resist the effects of weather and do not discolor under covering paint. Galvanized nails are available in many styles including the common and the finishing. When doing outdoor projects you use these since they are protected against rust *and* the discoloration that would result.

A very special nail is the scaffold nail we mentioned. This is often called a duplex or a double head. And that describes it exactly; the nail has two heads. You drive it to the first head for securing and use the second head to remove it. Why? It's designed for use on temporary structures.

Threaded nails are becoming more and more popular. These are screw-like in the way they grip and so have greater holding power. As the sketch shows, they are made to fit almost every building need.

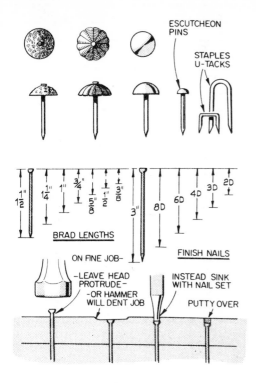

NAILING TECHNIQUES

Driving a nail should never be a question of showing your power. When wood fibers are torn and distorted as they are by excessively heavy blows, they do not grip the nail as they should. Hit firmly, yes, but not as you would in a carnival where you are trying for a prize. This gives you better control over the hammer anyway, with less chance of slipping off the nail-head and damaging the wood. Also, you feel less arm strain.

AVOID SPLITTING WOOD

Some woods have more tendency to split than others, especially when you are working along an edge or near an end. You can help to avoid this by dulling the point on the nail by tapping it gently with the hammer.

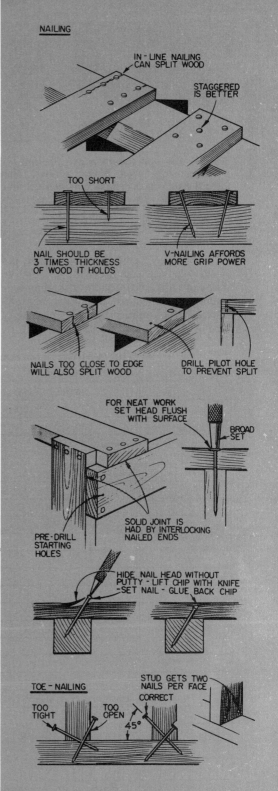

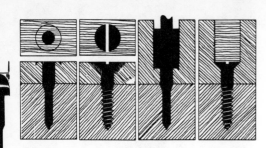

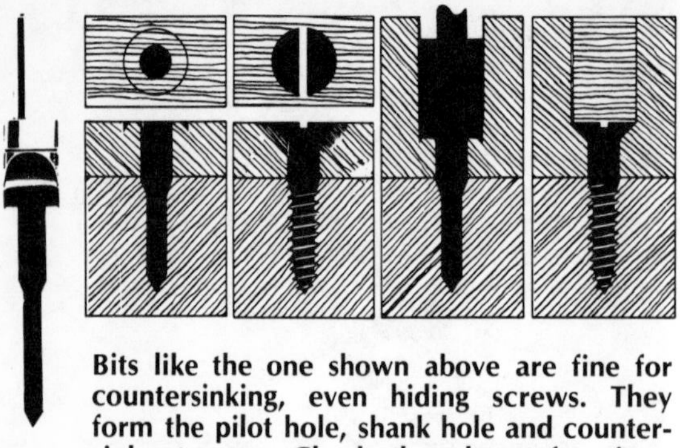

Bits like the one shown above are fine for countersinking, even hiding screws. They form the pilot hole, shank hole and countersink at once. Check the charts for sizes.

SCREWS

Screws have great holding power, pull parts together and are easy to remove. They are also more decorative than nails and many times are visually acceptable where nails would not be. Using nails to attach hinges would be a pretty sad touch and the hinge would loosen.

Doing a good job with screws is a

SCREW SIZES AND LENGTHS

Number of Screw	Shank Diameter	Lengths Available
0	.060	¼-⅜
1	.073	¼-½
2	.086	¼-¾
3	.099	¼-1
4	.112	¼-1½
5	.125	⅜-1½
6	.138	⅜-2½
7	.151	⅜-2½
8	.164	⅜-3
9	.177	½-3
10	.190	½-3½
11	.203	⅝-3½
12	.216	⅝-4
14	.242	¾-5
16	.268	1-5
18	.294	1¼-5
20	.320	1½-5
24	.372	3-5

PILOT AND SHANK HOLE SIZES

Screw Size	PILOT HOLE				SHANK HOLE	
	HARDWOOD		SOFTWOOD			
	Fractional Size*	Gauge Size	Fractional Size*	Gauge Size	Fractional Size*	Gauge Size
0	¹⁄₃₂	66	¹⁄₆₄	75	¹⁄₁₆	52
1		57	¹⁄₃₂	71	⁵⁄₆₄	47
2		54	¹⁄₃₂	65	³⁄₃₂	42
3	¹⁄₁₆	53	³⁄₆₄	58	⁷⁄₆₄	37
4	¹⁄₁₆	51	³⁄₆₄	55	⁷⁄₆₄	32
5	⁵⁄₆₄	47	¹⁄₁₆	53	⅛	30
6		44	¹⁄₁₆	52	⁹⁄₆₄	27
7		39	¹⁄₁₆	51	⁵⁄₃₂	22
8	⁷⁄₆₄	35	⁵⁄₆₄	48	¹¹⁄₆₄	18
9	⁷⁄₆₄	33	⁵⁄₆₄	45	³⁄₁₆	14
10	⅛	31	³⁄₃₂	43	³⁄₁₆	10
11		29	³⁄₃₂	40	¹³⁄₆₄	4
12		25	⁷⁄₆₄	38	⁷⁄₃₂	2
14	³⁄₁₆	14	⁷⁄₆₄	32	¼	D
16		10	⁹⁄₆₄	29	¹⁷⁄₆₄	I
18	¹³⁄₆₄	6	⁹⁄₆₄	26	¹⁹⁄₆₄	N
20	⁷⁄₃₂	3	¹¹⁄₆₄	19	²¹⁄₆₄	P
24	¼	D	³⁄₁₆	15	⅜	V

* Fractional size is approximate

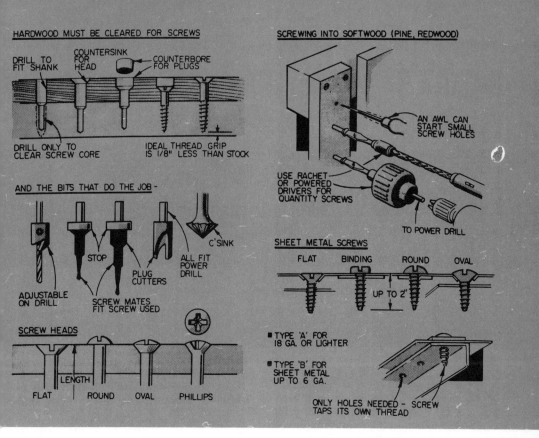

HARDWOOD MUST BE CLEARED FOR SCREWS

DRILL TO FIT SHANK
COUNTERSINK FOR HEAD
COUNTERBORE FOR PLUGS

DRILL ONLY TO CLEAR SCREW CORE
IDEAL THREAD GRIP IS 1/8" LESS THAN STOCK

AND THE BITS THAT DO THE JOB -

ADJUSTABLE ON DRILL
STOP
PLUG CUTTERS
SCREW MATES FIT SCREW USED
C'SINK
ALL FIT POWER DRILL

SCREW HEADS

LENGTH
FLAT   ROUND   OVAL   PHILLIPS

SCREWING INTO SOFTWOOD (PINE, REDWOOD)

AN AWL CAN START SMALL SCREW HOLES

USE RACHET OR POWERED DRIVERS FOR QUANTITY SCREWS

TO POWER DRILL

SHEET METAL SCREWS

FLAT   BINDING   ROUND   OVAL

UP TO 2'

■ TYPE 'A' FOR 18 GA. OR LIGHTER

■ TYPE 'B' FOR SHEET METAL UP TO 6 GA.

ONLY HOLES NEEDED - SCREW TAPS ITS OWN THREAD

question of following simple procedures. When the wood is soft and the screw is small, use an awl to create a starting hole. Doing this with the first screw on the job will tell you if it will work. If you have trouble driving the screw or can't get it down far enough for the head to seat correctly without destroying the threads in the wood, then you have to drill a starting hole and probably even form the countersink. On hardwoods, with large screws, you may even have to provide the shank hole. The size of the pilot hole and of the shank hole is fairly critical if you want the most holding power from the screw. But this is just a question of checking the chart for the size screw you are using. It's pretty bad policy to start a screw by driving it with a hammer.

SPECIAL SCREW BITS

Special bits are available that can be used in a drill press or in a portable drill.

These are made for specific size screws and will form the correct pilot hole, shank hole and countersink or counterbore. When you have many similar screws to drive, an item like this will make the job much eaiser.

Screws, like some nails, can be hidden, but it's not a good idea to use the putty technique. Standard procedure is to drive the screw through a counterbore-hole. This hole is then plugged with a dowel. Using regular dowel for the job may not matter on some projects but suppose you are working on mahogany? Then you buy a plug cutter so you can form the plugs from the same material. These are not expensive and lead to better craftsmanship since, by working carefully, you can cut the plug to match.

Like nails, screws are available in different materials and sizes. You can get them, for example, in brass or aluminum or you can get them coated for use where they might rust or become discolored.

NUTS, BOLTS AND OTHER FASTENERS

A knowledge of these items will separate the men from the boys

Skeleton framing for a built-in is being secured to house frame using a lag screw. These provide more strength than nails or ordinary screws. Drill pilot holes first.

Most common types of fasteners—in addition to nails and screws—are nuts and bolt combinations and lag screws. Nuts and bolts, even though you can buy them in small sizes, are considered heavy duty fasteners and most times you use them when you are not going through a conventional glue-joint procedure. You might assemble a piece of outdoor furniture with nuts and bolts or secure steel plates to exposed beams. You might use them on a workbench assembly or a stand you are making for a power tool.

MACHINE AND CARRIAGE BOLTS

Know the difference between a machine bolt and a carriage bolt. The machine bolt must be kept from turning as you take up on the nut. This means using a tool at each end and the necessity of keeping each end accessible should retightening be needed later on. A carriage bolt has a tapered shoulder under its head. This is forced into the wood as you take up on the nut and so keeps the bolt from turning. It means you can tighten or retighten without getting at the

head. Both of these come in galvanized versions for use outdoors.

LAG SCREWS

Lag screws are a combination of bolt and screw; you drive them like screws, provide pilot holes as you would for a screw, but you turn them with a wrench. Use these where you want more oomph than a conventional screw would provide —on such jobs as securing large hinges for heavy gates outdoors or when attaching steel rail to a porch.

Quite often they are used in combination with expansion sleeves when something must be attached to a masonry wall. The expansion sleeves are slipped into holes drilled in the masonry. When you take up on the lag screw, the sleeve expands to grip the walls of the hole. When using this idea in brick walls try to locate the hole for the expansion sleeve in mortar joints rather than in the brick.

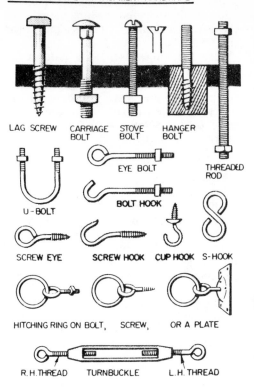

LAG SCREW   CARRIAGE BOLT   STOVE BOLT   HANGER BOLT

EYE BOLT   THREADED ROD

U - BOLT   BOLT HOOK

SCREW EYE   SCREW HOOK   CUP HOOK   S-HOOK

HITCHING RING ON BOLT,   SCREW,   OR A PLATE

R.H.THREAD   TURNBUCKLE   L.H.THREAD

MOLLY BOLTS

There are some types of fasteners for use on hollow walls where driving a nail or a conventional screw is impractical. Most popular type is the "Molly". This calls for drilling a hole to suit the size Molly being used. You then insert the fastener and take up on the attached screw. The wings of the Molly spread on the other side of the wall.

JACK AND TEE NUTS

"Jack-nuts" do a similar job and come in sizes that will grip material from "0" to 3/8" thick. Great, for example, for hanging a mirror on a hollow core door.

"Tee-Nuts" can be used to provide metal threads in plywood or lumber.

One piece of hardware along these lines we should mention that is practical for the homecraftsman is threaded rod which you can buy in various diameters.

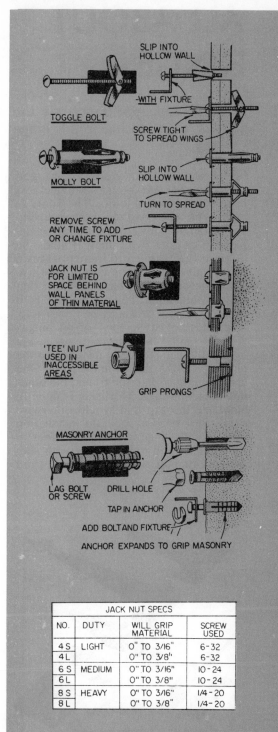

TOGGLE BOLT

SLIP INTO HOLLOW WALL

WITH FIXTURE

SCREW TIGHT TO SPREAD WINGS

MOLLY BOLT

SLIP INTO HOLLOW WALL

TURN TO SPREAD

REMOVE SCREW ANY TIME TO ADD OR CHANGE FIXTURE

JACK NUT IS FOR LIMITED SPACE BEHIND WALL PANELS OF THIN MATERIAL

'TEE' NUT USED IN INACCESSIBLE AREAS

GRIP PRONGS

MASONRY ANCHOR

LAG BOLT OR SCREW   DRILL HOLE

TAP IN ANCHOR

ADD BOLT AND FIXTURE

ANCHOR EXPANDS TO GRIP MASONRY

JACK NUT SPECS			
NO.	DUTY	WILL GRIP MATERIAL	SCREW USED
4 S	LIGHT	0" TO 3/16"	6-32
4 L		0" TO 3/8"	6-32
6 S	MEDIUM	0" TO 3/16"	10-24
6 L		0" TO 3/8"	10-24
8 S	HEAVY	0" TO 3/16"	1/4-20
8 L		0" TO 3/8"	1/4-20

ALL ABOUT HINGES

Strength and appearance are the factors in choosing right hinge

The pages of this book are hinged along one edge. Your knee and elbow have natural hinged joints. The doors in your home open and close through similar action but they employ a mechanical device that does two things. It holds the door to the frame and incorporates a pivot that allows the swinging action. The simple butt hinge has two leaves that wrap around a pin. The pin may be fixed in the assembly or it may be removable. Regardless of style, 99% of all hinges work in similar fashion.

A butt hinge is usually installed so the pin loops are visible from the front. One leaf is mortised into the frame, the other into the door edge. Mortising is done to eliminate the gap that would otherwise exist. Best bet for accurate installation of most any hinge is to put the door in the opening and hold it there with slim wedges or pieces of cardboard; then mark hinge locations on both door and frame. A door must always have clearance on all edges but the hinged one, and the cardboard shims can be a gauge for this.

Use the hinge itself as a pattern for marking the mortise. Mortise depth should be no more, no less, than the thickness of one leaf. You can cut the mortise with a portable router if you have one—or—outline the area with a sharp knife and then clean away waste with a chisel.

OFFSET HINGES

Offset hinges are used to mount lipped doors. Be careful when cutting the rabbet in the door edge since this should match the amount of offset in the hinge. Have the hinges on hand and, if checking is necessary, make a trial cut in some scrap wood before working on the doors. Offset hinges are available for surface-mounting or for semi-concealment.

DESIGN HINGES

A hinge can be purely functional—where you make every attempt to conceal it—or it can be surface-mounted to provide design detail. Oriental pieces with heavy brass hinges and escutcheons are good examples of design that rely heavily on fancy hardware. Wrought iron varieties for indoor or outdoor use also fall in this catagory. It's usually much easier to install the exposed, surface-mounted hinges than to work with concealed types—but the style of the project itself must be the deciding factor.

STRAP AND T-HINGES

Strap hinges and T-hinges are usually surface-mounted and are almost always used in utility or heavy duty applications. This doesn't mean they can't be decorative; they are available in many sizes and styles, but the common galvanized types are what you would use, for example, on a gate leading to a utility area or the door to a wood shed and that kind of thing.

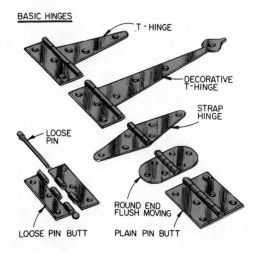

BASIC HINGES

T - HINGE

DECORATIVE T-HINGE

STRAP HINGE

LOOSE PIN

ROUND END FLUSH MOVING

LOOSE PIN BUTT

PLAIN PIN BUTT

MORTISING FOR BUTT HINGES

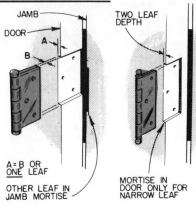

JAMB

DOOR

A

B

TWO LEAF
DEPTH

A = B OR
ONE LEAF

OTHER LEAF IN
JAMB MORTISE

MORTISE IN
DOOR ONLY FOR
NARROW LEAF

HINGES FOR INSET AND DROP DOORS

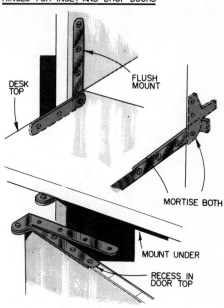

DESK
TOP

FLUSH
MOUNT

MORTISE BOTH

MOUNT UNDER

RECESS IN
DOOR TOP

FOR MORE ELEGANCE

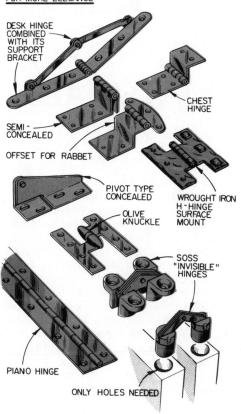

DESK HINGE
COMBINED
WITH ITS
SUPPORT
BRACKET

SEMI-
CONCEALED

OFFSET FOR RABBET

CHEST
HINGE

PIVOT TYPE
CONCEALED

OLIVE
KNUCKLE

WROUGHT IRON
H-HINGE
SURFACE
MOUNT

SOSS
"INVISIBLE"
HINGES

PIANO HINGE

ONLY HOLES NEEDED

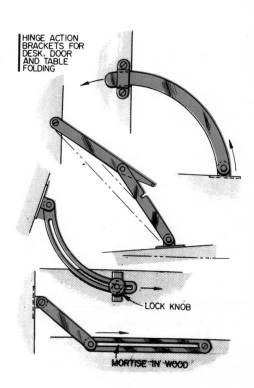

HINGE ACTION
BRACKETS FOR
DESK, DOOR
AND TABLE
FOLDING

LOCK KNOB

MORTISE IN WOOD

This type of paneling mounts on strips and makes it easy to decorate. Shelf brackets, picture hooks and similar items snap into the strips wherever you wish to place them.

KNOW YOUR MATERIALS

Send for catalogs, tour your "lumber yard" to know what's available

We think of materials as being those items you need to maintain, enhance or modernize your home. All supplies in these areas available to professionals are available to you. The local "lumber yard" is passé; today it's a homecraftsman's or "do-it-yourselfer's" supply center where you can buy anything from a washer to a pre-fab garden shed. It's almost routine at our house to walk through these places before starting any project, just to see what's available that will make the job easier to do and better to look at. Many times, the sightseeing leads *to* a project. Once we were able to build an inside brick wall without pouring a concrete foundation because we discovered plastic brick light enough so we could eliminate a footing.

All craftsmen should maintain a file of manufacturer's catalogs. These will keep you up to date on materials and will also tell facts concerning their use and installation. Getting these catalogs is simple—just look through do-it-yourself books and magazines and send away to manufacturers for their literature. We can't tell you all there is to know here; this book or a book twice its size is too small. Also, we would just be repeating facts about specific products already told in greater detail in the maker's manuals.

SOME THOUGHTS ON LUMBER

Wood is either "hard" or "soft", but

WOOD	DESCRIPTION	TYPICAL PROJECTS
ASH	Hardwood, heavy, with good strength and open grain, that takes gluing fairly well and is easy to nail—good for turning.	Ball bats, frames for boats, handles for tools, paneling veneer
BIRCH	Hardwood, heavy, with great strength that glues fairly well, but nails poorly. Has close grain and is good for turning. Often mistaken for Maple.	Bowls and plates, furniture, dowels, interior trim
CHERRY	Hardwood, medium weight, with fair strength and a grain that can stand filling. Excellent for gluing and turning.	Cabinet making, furniture, turned projects, handles
GUM	Rates about the middle in hardness, weight, strength and grain. Very good for gluing, nailing and turning. Often a hypocrite wood for walnut.	Use as a substitute wood for walnut—most always stained to resemble something else.
MAHOGANY	Medium hardwood, heavy, with fair strength and open grain. Very good for gluing, nailing and turning. Honduras best, Phillipine cheapest.	Fine mahogany for cabinets, furniture, boats, veneers—others for plywood facings.
MAPLE	Heavy hardwood, strong, with close grain that glues and turns pretty good but is tough to nail. Some species just a little softer.	Good furniture wood, popular for colonial designs, flooring, handles, bowls, wooden ware
OAK	Heavy hardwood with great strength and very open grain that's about the middle of the road for gluing, nailing and turning.	Heavy-duty furniture, boat frames, desks, handles
PINE (harder)	Species of pines that are heavy, strong and semi-hard with fairly open grain. Glues fairly well but is poor for nailing and turning.	Functional applications, some interior finish stock
PINE (softer)	Like the sugar pines that are light in weight, not too strong, soft, close grained. Great for gluing, nailing, and not too bad for turning.	Trim stock, moldings, window and door stock. Functional shelves
REDWOOD	Lightweight wood, soft, with close grain and good strength. Good for gluing, nailing, turning. Redwood burl makes beautiful turnings.	House covering and trim, outdoor furniture, fences, planters, interior wall covering, inside trim
WALNUT	Classified as hardwood of medium weight, strength and grain. Good for gluing and nailing and excellent for turning.	Quality wood for furniture, wall paneling, turnings, gunstocks, cabinetry, novelties

don't take the terms literally; they are botanical catagories that indicate the wood has come from a broad-leafed, deciduous tree (hardwood) or a cone-bearing or evergreen tree (softwood).

Maple, birch, mahogany, walnut, oak are typical hardwoods. Cedar, pine, fir, redwood, are typical softwoods. If you have ever worked with mahogany you know that it isn't hard in the sense that it's difficult to cut. Fir is a softwood but not in the sense that you can work it easily with a kitchen knife.

Wood can be "open-grained" or "close-grained". This relates to the cellular structure of the species. Oak is a good example of open-grain wood while maple is a typical species of close-grained woods. This characteristic affects finishing procedures. Open-grained woods require a filler to pack the pores so the finish will be smooth. Close-grained woods do not need filling since a good sanding job produces a smooth finish.

A tree gets bigger by adding layers of wood each year. These "growth rings", which you can count to determine the age of the tree, are what contribute to the

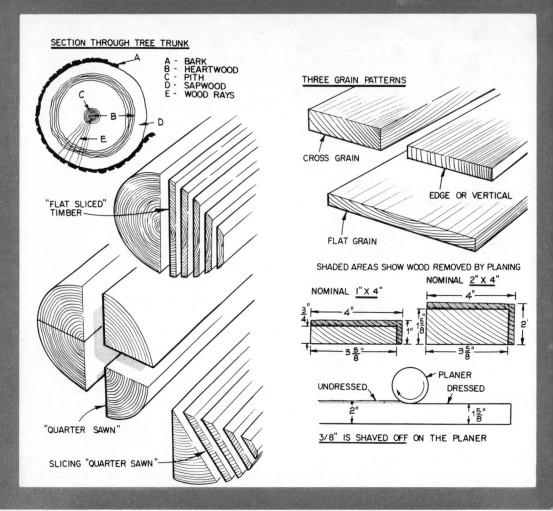

SECTION THROUGH TREE TRUNK

A - BARK
B - HEARTWOOD
C - PITH
D - SAPWOOD
E - WOOD RAYS

"FLAT SLICED" TIMBER

"QUARTER SAWN"

SLICING "QUARTER SAWN"

THREE GRAIN PATTERNS

CROSS GRAIN

EDGE OR VERTICAL

FLAT GRAIN

SHADED AREAS SHOW WOOD REMOVED BY PLANING

NOMINAL 1" X 4"

NOMINAL 2" X 4"

PLANER

UNDRESSED

DRESSED

3/8" IS SHAVED OFF ON THE PLANER

grain pattern when the tree is sliced into boards. This pattern can vary from tree to tree in the same species and even from area to area in the same tree but it is still characteristic enough so you can recognize a particular species. How the tree is cut up also effects the grain pattern in the board.

SLICING LOGS

Slicing lengthwise produces wide boards with prominent grain. Milling this way is economical and most yard lumber is produced in this fashion. Quarter sawing is like cutting the log, initially, into four sections that look like giant pieces of quarter-round molding. The broad sides of the "molding" are then sliced. An even, attractive grain pattern results but since it's an expensive milling technique it's used mostly to get boards from some of the more rare hardwoods.

ROTARY CUTTING

A log can be placed in a giant lathe-like machine and then turned against a knife so long, thin sheets of veneer are peeled off. This is rotary cutting. To make a common type of plywood, an odd number of veneers are placed at right angles to each other and glued together.

BUILDING LUMBER—UTILITY BOARD GRADES

#1 common	adequate for general-purpose work—contains small, tight knots—use on outside trim, door and window frames, siding	#5 common	bottom grade and seldom shipped out of mill locality
#2 common	contains knots and some may not be sound—use for sub-flooring, some siding, "hidden shelves"	B Select and better	tops—hard to find a blemish—use for built-ins, interior trim, cabinet work, furniture
#3 common	more knots and many unsound—also pitch pockets and stains—can be culled for shelving use	C Select	imperfections may be easily covered with paint—can be used as B Select but inspect carefully first
#4 common	worse than #3 so not too useful except for temporary structures, boxes, crates	D Select	low man on the better grade pole—imperfections similar to those in C Select but more of them—use for painted interior work, shelves, etc.

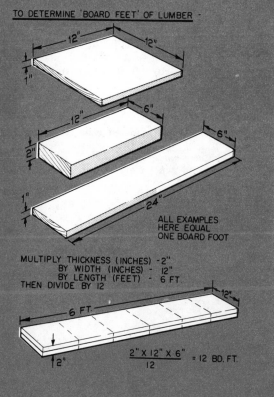

TO DETERMINE 'BOARD FEET' OF LUMBER -

12"
12"
1"

12"
6"
2"
6"

1"
24"

ALL EXAMPLES
HERE EQUAL
ONE BOARD FOOT

MULTIPLY THICKNESS (INCHES) - 2"
BY WIDTH (INCHES) - 12"
BY LENGTH (FEET) - 6 FT.
THEN DIVIDE BY 12

6 FT.
12"
2"

$$\frac{2" \times 12" \times 6"}{12} = 12 \text{ BD. FT.}$$

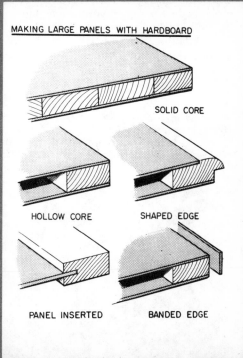

MAKING LARGE PANELS WITH HARDBOARD

SOLID CORE

HOLLOW CORE

SHAPED EDGE

PANEL INSERTED

BANDED EDGE

Outdoor projects call for materials that will hold up in weather. Redwood and cedar are typical woods used. If gluing is involved, make sure it's waterproof glue.

For delux indoor projects, use the fancier hardwoods. Since a natural finish is almost standard, such jobs call for more care and craftsmanship than applied to painted jobs.

There are other methods of producing veneers; some of them quite similar to the cutting techniques used to get boards. Each produces a particular grain pattern. You can buy veneers to glue to your projects or to make your own panels. Beautiful slabs (for table-top use, for example) can be produced at home by following the grain-matching techniques shown in the drawings. You don't have to make the entire panel—just glue the veneer to a plywood panel you buy.

DRYING LUMBER

Wood, when first cut, contains a considerable amount of moisture. If used as is, or allowed to dry in an uncontrolled

When a project is going to take a beating, think in terms of durable products. Marlite was used here, both on top and door fronts. Other parts are plywood, stained to match.

If you crave oak paneled walls, relax. You can put it up in big 4×8-ft. sheets. A Royalcote product by Masonite is factory-finished and wipes clean with a damp cloth.

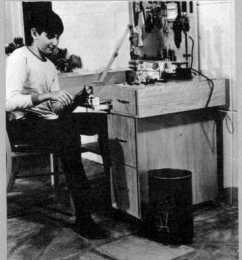

Modern plywood products include items like this fretwork piece that requires nothing but framing to create an attractive divider or screen. Can also be used for doors.

You can do your own fancy work with some of the veneers available if you have the means of pressing them together with glue. Screws you buy, press you make yourself.

fashion, considerable distortion may result. This is because the moisture content and the wood structure is not uniform. Since you don't want this to occur *after* you have made a project, the moisture content of the wood is reduced to an acceptable minimum before it gets to you.

Mostly, it's done by placing the green wood in storage sheds where the atmosphere can be controled. Steam may be introduced so the moisture content in the wood will become as uniform as possible. This minimizes distortion when the wood is finally dried.

Air-drying is simply a way of stacking wood carefully with spacers between to provide adequate ventilation .

Acoustical tile comes in different sizes and textures. This is a suspended ceiling. Tiles rest on gridwork that hangs from a series of wires that are nailed to joists.

A new floor, easy to put down when working with tile, can change the character of the average room—and quickly. Products like this Armstrong Excelon are easy to install.

HINTS ON BUYING AND USING

Don't, as we said, buy enough to start your own little lumber yard. Get what you need for the project—let the dealer worry about storage space and possible distortion.

Always buy the cheapest grade that is adequate for the job.

When you plan a small project but want a deluxe finish, you can often save money by buying a lower grade of material and culling out the good sections. For example, by buying a decent grade of pine shelving, you can produce a good amount of clear pine merely by cutting out the knots. Naturally this can't always work. But when you figure pine shelving at 10 to 12 cents a foot and compare it to clear stock at up to 60 cents a foot, you can see that the idea might pay off occasionally.

ACCLIMATE AND SEAL

Whatever wood material you buy, store it for a bit in the area where it will be used. This will let it get accustomed to the new atmosphere so possible slight changes will happen *before* not *after*. This applies to lumber, shaped lumber paneling, plywood panels—even pre-finished varieties, acoustical tile, etc. Actually, any material that is porous enough to absorb moisture.

Finishing is important, not just for looks, but for protection. No matter what the project is, even one made from cheap lumber that won't be painted, put a coat of sealer on it.

HOW TO ORDER

Most times, when you order, it pays to specify the lengths you want. Don't ask for X number of feet. Instead, for example, order 8 pieces, 10 feet long or 3 pieces 12 feet long. This, of course, after you have judged your requirements and made a logical decision on most suitable lengths. *Don't*, however, ask for specific sizes for the project; stick to stock sizes. Special cutting by the lumber yard costs money.

PLYWOOD

Plywood is basically a sandwhich of veneers. It is, however, available with a solid lumber core, a particleboard core, even a hardboard core. The solid core examples are usually more expensive and often are faced with fancy wood veneers.

Plywood is practical since it gives you large, stable panels, ready to work with. In effect, since the surface coating can be a thin veneer, it makes available greater quantities of rare woods.

The most practical panel for shop use and for many projects is made of fir and is available in 4' × 8' sheets. Larger sizes are available—5' × 9' for a ping-pong table for example, but these are usually made available on order.

TYPES OF PLYWOOD

There are many types. The best way to ask for it is to identify it with the project you are planning. There is plywood for underlayment, sheathing, sub-roofing, concrete forms, common grades of shop plywood that you can cut up for shelves and so on. Fir plywood is often used for inside projects but these call for the higher, sanded grades that are free of of blemishes on one or both faces.

Common thicknesses are ¼", ½" and ¾" but ⅜" and ⅝" can also be obtained. A special 1-⅛" panel, matched on all four edges for continuous runs in any direction is available for sub-flooring over post and beam construction.

HARDBOARDS

Hardboards are also a wood product and have come a long way since they were first introduced. At one time they were thought of as utility material for drawer bottoms and cabinet dividers and so on. Today they are used for wall paneling—as is, or coated with wood-grained plastics—and can be purchased embossed or perforated. Marlite is an example of plastic-coated hardboard that is available in sheets for cutting or for use as wall covering. Special types are avail-

WORKING WITH THIN VENEERS - THESE ARE EXAMPLES OF SEGMENT MATCHING YOU CAN DO

PLYWOOD COMES FROM SOLID WOOD TIMBER SHAVED ALONG GRAIN BY ROTATING HUGE LOGS

THEN BONDED INTO LAYERS, IMPARTING GREAT STRENGTH TO LARGE PANELS

HEAVY PLYWOOD BASE

LONG VENEERS FORM BORDERS

EDGE VENEER

POPULAR THICKNESS OF COMMON PLYWOOD IN 4" X 8" PANEL SIZES

EACH SIZE ALSO COMES IN A WIDE VARIETY OF FINE WOOD VENEER FINISH SURFACES

able that stand up in moisture areas. You can, by using special extrusions and caulking, even use it in a shower.

Perforated hardboards are great for storage walls. Special pieces of hardware are available so you can hang most anything. They can even be used for acoustical purposes when backed up with a sound-absorbing material—something to consider when covering shop walls.

KNOW WHAT'S AVAILABLE

If you check through catalogs and local supply houses you'll find many materials that can be put to good use by the homecraftsman. Cast or carved plaques that can be used to make fancy doors or lend an exotic touch to a project, pre-carved brackets for shelves, ready-made furniture legs in metal or wood, self-adhesive banding for covering plywood edges, moldings in endless variety, perforated or embossed plastic and wood-product sheets that are great for dividers and screens, and on and on. Checking it all out will not only make things easier for you but could very well inspire you to accomplish great things with minimum effort and in minimum time.

As illustrated here by Simpson, door styles have changed along with other things. Quite often you can give a room, or an entranceway, a new look by merely replacing the door.

HOW TO HANG DOORS

Here is how to dress sticking doors and how to hang new assemblies

The difference between the entrance door of your home and the inside doors is in weight and width. Entrance doors are usually solid and three feet wide. Because of this, they hang, or they should hang, on three hinges. Interior doors are narrower and usually of hollow-core construction which makes them light enough so two hinges will support them adequately. Three hinges may be used, not for support, but to provide an extra guard against distortion.

Probably the only maintenance job you'll ever have to do on a door is to dress it should it begin to stick. This can be caused by some house-settling, by swelling of the door because it wasn't sealed correctly when installed, because adequate clearance between door-edge and frame was not provided, or because hinges have loosened.

TIGHTEN SCREWS

When sticking does occur, take up on all the hinge screws until they are firmly

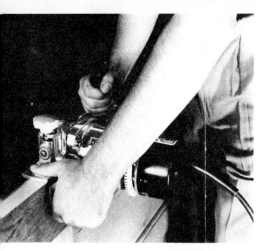

If you own a portable router, chances are you can buy a planer attachment for it; a fine way to fit doors or shave them down to ease binding. Hand plane can be used.

When fitting a door to an existing frame, dress it to fit the opening and hold it in place with shims as you mark locations for hinges. Leave 1/16" clearance along edges.

When installing your own hardware, work carefully with the templates supplied for the purpose. These will tell you where to drill and how large the holes should be.

If you have many doors to do and plan to cut your own mortises, rent an outfit for the job. This includes a router and a template. Use template on all doors needed.

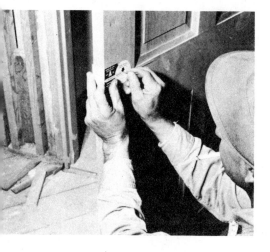

seated. An extreme problem here would be a loose hinge that can't be retightened with the original screws. Some simple techniques often provide a cure for this. Try forcing a splinter of wood or a wooden matchstick in the holes; see if this won't cause the screws to grip. If it doesn't, try the next size screw. This may make it necessary to deepen the countersink in the hinge just a bit. If this doesn't provide enough security, you can try drilling an extra hole or two through the hinge so you can add another screw.

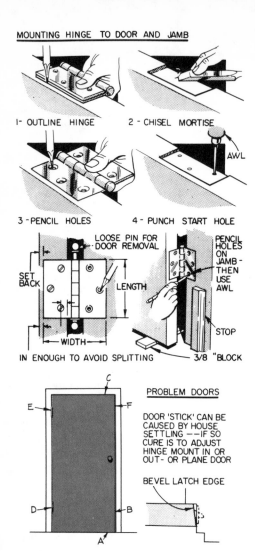

MOUNTING HINGE TO DOOR AND JAMB

1 - OUTLINE HINGE   2 - CHISEL MORTISE

AWL

3 - PENCIL HOLES   4 - PUNCH START HOLE

LOOSE PIN FOR DOOR REMOVAL

PENCIL HOLES ON JAMB - THEN USE AWL

SET BACK

LENGTH

STOP

WIDTH

IN ENOUGH TO AVOID SPLITTING   3/8 "BLOCK

PROBLEM DOORS

DOOR 'STICK' CAN BE CAUSED BY HOUSE SETTLING ──IF SO CURE IS TO ADJUST HINGE MOUNT IN OR OUT- OR PLANE DOOR

BEVEL LATCH EDGE

If door sticks at A or F, check E hinge for tightness. If door sticks at C try a thin shim at E. If door rubs at B and the hinges are tight, use plane or sandpaper to reduce high spots. If door scrapes at A—shim the door out at hinge marked D.

If the screw-hole area is damaged beyond any repair of this nature, you'll have to go further.

If the damage is in the door, you can, of course, replace it or you can cut out the bad part and glue in a new block. This requires careful cutting so you don't leave an obvious flaw. If the damage is

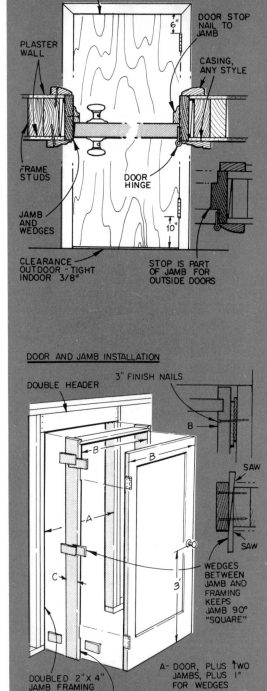

CLEARANCE TO PASS 5¢ COIN

DOOR STOP NAIL TO JAMB

6"

PLASTER WALL

CASING, ANY STYLE

FRAME STUDS

DOOR HINGE

JAMB AND WEDGES

10"

CLEARANCE OUTDOOR - TIGHT INDOOR 3/8"

STOP IS PART OF JAMB FOR OUTSIDE DOORS

DOOR AND JAMB INSTALLATION

DOUBLE HEADER

3" FINISH NAILS

B

B

B

SAW

A

SAW

C

WEDGES BETWEEN JAMB AND FRAMING KEEPS JAMB 90° "SQUARE"

3'

DOUBLED 2"X 4" JAMB FRAMING

1 1/8"STOCK JAMB

A- DOOR, PLUS TWO JAMBS, PLUS 1" FOR WEDGES
B- DOOR WIDTH
C- STUD WIDTH PLUS WALL PANELS

in the jamb, you can replace that or do a similar repair job. Actually, such a repair job on a jamb can be more frustrating than the replacement chore.

SHIM HINGES

If you have determined that the problem is not caused by hinge looseness, then check the following procedures. If the door sticks at the bottom on the knob side and you have adequate clearance at the top and knob edge, try a thin cardboard shim (at the bottom) between jamb and hinge. If clearance doesn't permit this, place a piece of sandpaper between the floor and the bottom edge of the door and work the door back and forth over it. Don't use a very coarse paper since this can splinter surface-veneers.

If the door sticks at the top edge and you have ample clearance at the bottom, try the shim idea under the top hinge.

SAND OR PLANE

If these ideas don't work for you, then the job must be done by removing material in the binding areas. Usually, a stroke or two with a block plane is enough to do the job. Sometimes you can accomplish it merely by rubbing with sandpaper wrapped around a wood block. Remember, however, that the door does require clearance at the top, the bottom and the knob edge. If this means planing away 1/16" of material, then do it. Also remember that removing material removes the original seal coat. Thus you must re-seal those places ·

REPLACING DOOR

Replacement of an entire door is accomplished most easily by buying a complete door assembly. This includes door *and* frame with the door already hinged in place and ready to receive remaining hardware. Even contractors work this way today since it's much simpler than having a carpenter do the framing and fitting at the site. To do a good job in this fashion, check the drawings in this sec-

Doors fold up as well as swing. Easiest installation hinges the door sections to each other. Outside section is hinged to jam. Large doors have a guide track on the top.

tion that show the rough opening into which you fit the finished door frame (jambs). Important points to remember are these—use tapered shims to establish frame position. Work carefully to get a square installation. Be sure the jambs are wide enough to match the wall thickness, and this includes the wall covering. Use casing or finishing nails to secure the door frame in the opening.

When you are replacing a door without removing the original frame, dress the door to fit before you do anything else. Once this is accomplished, put the door in place—holding it there with wedges. Then carefully mark the hinge locations on the new door. The drawings show how mortising for the hinge can be accomplished with hand tools.

Floating shelves have no visable means of support. Steel bars are forced into holes drilled into the studs. Predrilled shelves are pressed onto the bars for neat effect.

HOW TO CREATE SHELVING

Shelves are storage areas and should be designed for specific use

So what's with shelves? Aren't they all pretty much alike? Yes—in the average house—but that's what we want to get away from. It's not far-fetched to say that a series of equally-spaced, fixed shelves is poor utilization of the space they occupy. The only justification for that kind of thing, by way of example, would be several hundred books all of which are exactly the same size.

For the most part however, shelves are storage areas and the items to be stored seldom are the same shape and size. When you custom-design shelves for particular items you not only *utilize* space but you make it more convenient to store and use the items.

When you install shelves that are adjustable, you take a step in that direction since this makes it possible, at least, to plan for the average height of different storage items. It's a step in the right direction but it's still a compromise. Better to use triangle shelves or step shelves or

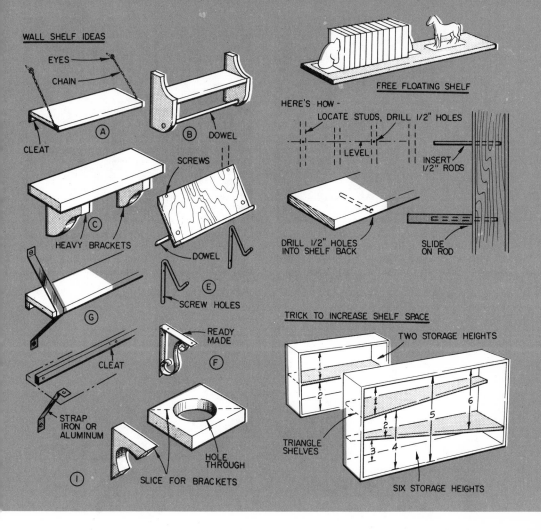

WALL SHELF IDEAS

EYES

CHAIN

CLEAT

(A)

(B) DOWEL

SCREWS

(C)

HEAVY BRACKETS

DOWEL

(E)

SCREW HOLES

(G)

READY MADE

CLEAT

(F)

STRAP IRON OR ALUMINUM

(I)

HOLE THROUGH

SLICE FOR BRACKETS

FREE FLOATING SHELF

HERE'S HOW -

LOCATE STUDS, DRILL 1/2" HOLES

LEVEL

INSERT 1/2" RODS

DRILL 1/2" HOLES INTO SHELF BACK

SLIDE ON ROD

TRICK TO INCREASE SHELF SPACE

TWO STORAGE HEIGHTS

2

TRIANGLE SHELVES

2

3

4

5

6

SIX STORAGE HEIGHTS

Some types of wall paneling mount to provide hanging strips for special pieces of hardware that can be placed anywhere for shelves, even for hidden picture hangers.

balcony shelves or a combination of two or all of them. Check the drawings and you will see how, by "remodeling" a single, fixed shelf to two triangle shelves can provide six storage heights instead of the original two.

GOOD DESIGN

Good shelf design is mostly a matter of getting away from the fixation that a shelf is a fixed, horizontal slab. Or that it must be as wide as the depth of the cabinet. Some wives get pretty shaken at the thought of having to reach blindly into a cavern to get to something at the

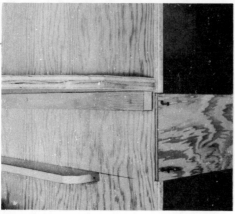

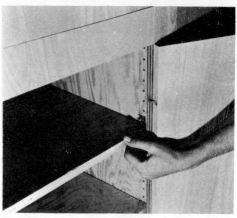

Very narrow balcony shelves to fill spaces in built-ins or cabinets will provide storage space for small items such as cordial glasses. Set on cleats or nail and glue.

Simple way to provide for a series of adjustable shelves is shown. Drill two lines of holes in case sides, use store-bought brackets than can be conveniently placed.

HOME

OFFICE OR SHOP

You can buy special shelf brackets like the wrought iron example shown that will make up into very suitable display shelves. Note that the bracket can be used two ways.

Fold-down shelves are possible by using a special set of hinges that also lock into supports. These can be used for fold-away breakfast bars, work areas, or whatever.

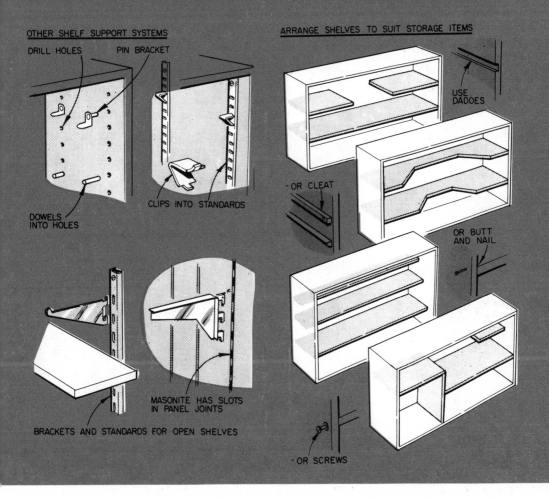

OTHER SHELF SUPPORT SYSTEMS

DRILL HOLES   PIN BRACKET

CLIPS INTO STANDARDS

DOWELS INTO HOLES

BRACKETS AND STANDARDS FOR OPEN SHELVES

MASONITE HAS SLOTS IN PANEL JOINTS

ARRANGE SHELVES TO SUIT STORAGE ITEMS

USE DADOES

- OR CLEAT

OR BUTT AND NAIL

- OR SCREWS

back of the shelf. Balcony shelves would solve this problem and the "empty" space in the center of the cabinet could be occupied by tall items that equally-spaced shelves would not provide for.

It takes some pre-planning and visualization but that's not a problem—you merely set out, measure and study the items you plan to store; then build .

SHELVES IN CLOSETS

Now this holds true for more than chinaware. The difference in length between a man's jacket and a woman's dress can be a factor in shelf height in a clothes closet. Shelf height in a child's closet should not be the same as a similar item in an adult's closet. Which, if you have already read the chapter on furniture, brings us back to the thought that good design must be based on function. In the case of the shelf in the child's closet you consider how high the child can reach. Go further by designing that shelf so it can be raised as the child grows up. This doesn't have to be anymore than an adjustable shelf idea.

In some instances, shelves have to be attractive as well as functional. This occurs mostly when the shelves are not hidden by doors and they can be put in the catagory of "display" shelves. The drawings and photos are also intended to lead you in this direction.

The idea, really, is to give the project some thought. A shelf can be—should be—more than a board sitting on cleats.

This closet provides for "his and her" dressing rooms. Note shelf and drawer install-ations, even mirror length, are designed for the user. Careful planning reaps rewards.

INSIDE STORY ON CLOSETS

Think of closet space as you would a built-in and use every inch

You can do much to increase storage facility in your home merely by re-designing the interior of closets already there. We're speaking primarily of cloth-ing closets; those caverns with a single shelf and a single clothes pole, and often with a single door so that space to either side is difficult to use.

STORAGE SPACE

Storage space, on the whole, is not meant for display. It's X number of cubic feet that should be utilized. If you place a hat on a shelf that has two feet of space above it, all the space above the hat will be wasted. You can stack the hats if you put them in boxes but this makes it in-convenient to get to a specific one. In es-sence, if you solve this problem by mak-ing a special compartment for each hat, you have approached the solution to uti-lization of available space. This, in itself, gives you more space simply because you use all that's there.

The carpentry involved should follow

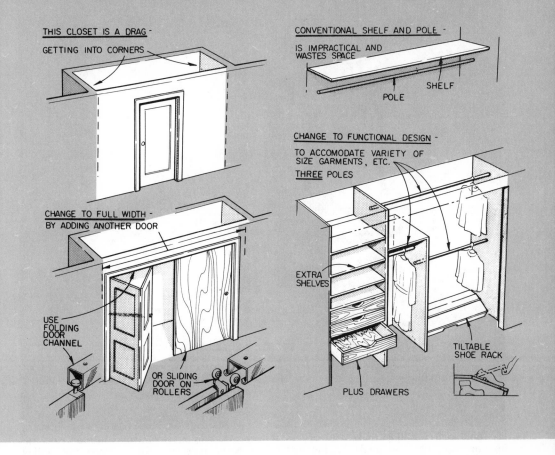

THIS CLOSET IS A DRAG -
GETTING INTO CORNERS

CONVENTIONAL SHELF AND POLE -
IS IMPRACTICAL AND
WASTES SPACE
SHELF
POLE

CHANGE TO FULL WIDTH -
BY ADDING ANOTHER DOOR

USE FOLDING DOOR CHANNEL

OR SLIDING DOOR ON ROLLERS

CHANGE TO FUNCTIONAL DESIGN -
TO ACCOMODATE VARIETY OF
SIZE GARMENTS, ETC.
THREE POLES

EXTRA SHELVES

PLUS DRAWERS

TILTABLE SHOE RACK

a period of planning which is not more than a matter of looking at the storage items. A man's coat and folded trousers require so much vertical space. In the average closet you can mount two poles so items of this nature could be stored one above the other. A topcoat requires more vertical space but chances are you have more suits than topcoats so storage for the latter can be provided by fitting a short pole between partitions. When you plan to provide mostly for the extremes, you usually waste space.

CONSTRUCTION

The actual construction should be viewed as if it were possible to pre-assemble a unit and then slide it into place in the closet. In an existing closet this can be done to a point by making sub-assemblies and then tying them together after they are in place. For tying to existing walls, just check the section on house framing. Stud, plate, fireblocks, headers, are exactly the same as they are for any other area. Once you locate one stud, you can measure off to other members for nailing purposes.

ADDING A CLOSET

Adding a closet is, again, a question of basic structural techniques. The drawing shows how you can build a closet in a corner working with the same materials you would use for the house skeleton. When possible, plan the depth and the length of the addition so you can tie in strongly to existing framing members. But this is not essential since the new structure will not have bearing walls.

Plywood wall paneling makes it easy to redecorate, and in worthwhile fashion. Just doing one wall of room makes a big difference. Note molding finish at floor and ceiling.

HOW TO PANEL WALLS

The new paneling is attractive, easy to install and maintenance-free

The most common type of wall paneling is plywood and it comes in 4×8 ft. sheets than can be from ⅛″ to ¼″ thick. The sense behind this makes available fancy, even rare woods, in a form easy to apply and at a price far below what you would have to pay for the same material in lumber form. Some types of mahogany paneling are sold at $6.00 to $8.00 a panel and this covers 32 square feet. If you were doing a wall 15 ft long you would need four panels. The price of these plus other materials needed would bring the cost of the whole job to not more than $40.00. And it would be a lifetime wall since the panels are factory-finished and can be maintained like new by dusting and waxing them like fine furniture.

NEW PANELING

New paneling can be applied to open studs, directly to an existing wall or on furring strips. Furring strips are used to create a new nailing surface for the panels and are especially needed when the wall (covered or open studs) is uneven, badly cracked or otherwise unsuitable for direct application.

80

OVER OPEN STUDS

When you work over open studs, you want to be sure to eliminate high spots. Check for these by stretching a line across the wall at various points and marking the high spots with chalk; then use a plane to reduce those areas. It's not likely that you will find very bad low spots but should you, fill them by using thin pieces of wood as shims. Studs that are real bad should be replaced with straight ones.

Panel joints should fall on a stud center. This means, assuming that you start from a corner with a full panel, a stud every four feet even if you must add additional ones.

FURRING STRIPS

The application of furring (over open studs or existing wall) is to create a "frame" for nailing purposes and to provide a solid feel for what might be a thin panel. Thus, furring is attached horizontally across the wall (see drawing) and vertically over studs on 4 ft. centers. There would be no objection should you choose to run vertical furring over *all* studs. All openings, doors and windows, must have perimeter furring.

Furring material doesn't have to be fancy but should be sound. A dry, shelving type of knotty pine works out okay. A 1×3 material is good although you can go wider if you wish. If you own a table saw, buy the wood in 12″ widths and rip to furring size in your own shop. This will save you some money. When attaching, use common or box nails that are long enough to go through furring and wall covering (if any) and penetrate the studs *at least* 1″. Use two nails at every stud-crossing and about a 12″ spacing on

Always start the first panel in a corner. Open edge must fall on a stud center even if you must reduce the panel width or add special stud. Study drawings for procedure.

Check open edge of first panel carefully for vertical alignment. All the others will be right — or — wrong, depending on how well you succeed in doing this one.

This type of wall paneling (by Masonite) hangs on perforated metal strips that are nailed to studs. Special fixtures hook to strips so you can hang shelves anywhere.

Cut openings carefully with a keyhole saw or saber saw. When you have many switches or outlets to cut for, it pays to make a cardboard pattern for marking the opening.

Finishing at doors is easy if you make special jambs to cover added wall thickness. Molding covers the opening between the jamb and the wall. See drawings for other ideas.

This wall was first covered with plasterboard so paneling would have a more solid backing. Adhesive was used for application so that no furring would be needed for job.

This Georgia-Pacific wall paneling product comes with matching moldings. Note that craftsman decided to use matching molding at ceiling—contrasting molding at floor.

Adhesives can be used to attach furring as well as paneling. Be sure to follow instructions that come with product. Where possible, use nails or fasteners with glue.

To use adhesive, squeeze out line at contact points, press on panel. Pull panel away for a second and return to original position. Tapping with a hammer will help.

Lumber-type paneling is done the same way except less furring is okay. Cut carefully around openings. The joint between window and paneling is then covered with trim.

all perimeter work. Use a level as you apply the furring and place thin pieces of wood behind the furring should you find low spots. Place shims over studs so furring nails will go through them.

START FROM CORNER

Always start the first panel from a corner. This one is important since it will establish vertical alignment of the others. The first panel must fit the corner along one edge and fall on a stud centerline at the other edge. If the adjacent wall is not plumb you can transfer its line to the panel by using a compass. If you plan to use molding in that corner, don't fret too much about the joint since it will be covered. This applies also if you plan to turn the corner with paneling.

Nail panels up as you go using the size and type nail recommended by the manufacturer of the panels. Some makers supply nails finished to match the panel tone. If ordinary finishing nails are suggested, be sure to use a nail set for the final driving—getting the head just a bit below the wood surface. Those holes will be filled later with a matching color putty stick.

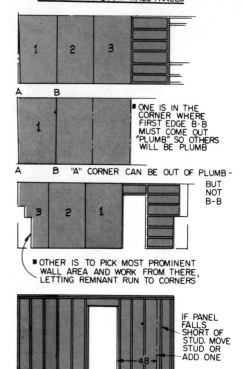

TWO WAYS TO START WALL PANELS

ONE IS IN THE CORNER WHERE FIRST EDGE B-B MUST COME OUT "PLUMB" SO OTHERS WILL BE PLUMB

"A" CORNER CAN BE OUT OF PLUMB - BUT NOT B-B

OTHER IS TO PICK MOST PROMINENT WALL AREA AND WORK FROM THERE, LETTING REMNANT RUN TO CORNERS

IF PANEL FALLS SHORT OF STUD. MOVE STUD OR ADD ONE

NEW OR OPEN STUD WALL IS OKAY FOR PANELS WITHOUT USING FURRING STRIPS, IF STUDS ARE SPACED CORRECTLY

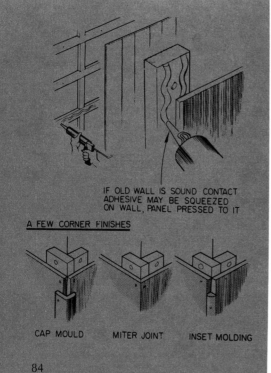

IF OLD WALL IS SOUND CONTACT ADHESIVE MAY BE SQUEEZED ON WALL, PANEL PRESSED TO IT

A FEW CORNER FINISHES

CAP MOULD   MITER JOINT   INSET MOLDING

USING MOLDING

If you determined beforehand that molding would be used at the floor, ceiling, etc., then joints at those places are not too critical. Without molding, you must do a more craftsman-like job. But this doesn't mean forcing the panel between floor and ceiling—this would do more harm than good. A slight gap is in order; maybe a 1/16th of an inch. The craftsmanship comes in getting that gap even all along the line.

Paneling can be attached with special adhesives. These vary from product to product so use should be in strict accordance with manufacturer's instructions. Most of them come in tubes and are squeezed out like weatherstripping from a caulking gun.

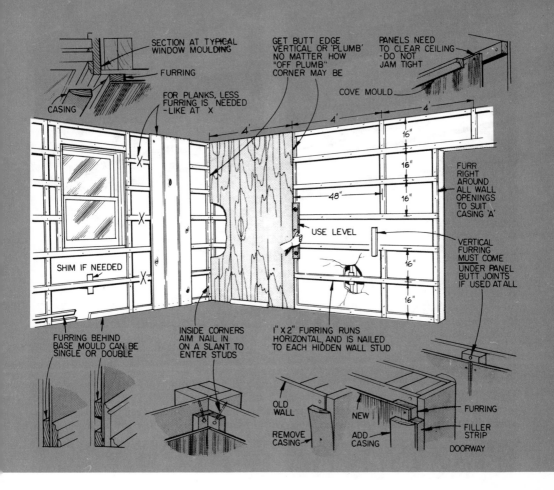

SECTION AT TYPICAL WINDOW MOULDING

FURRING

CASING

FOR PLANKS, LESS FURRING IS NEEDED –LIKE AT X

GET BUTT EDGE VERTICAL OR 'PLUMB' NO MATTER HOW "OFF PLUMB" CORNER MAY BE

PANELS NEED TO CLEAR CEILING –DO NOT JAM TIGHT

COVE MOULD

4'

4'

4'

16"

16"

48"

16"

FURR RIGHT AROUND ALL WALL OPENINGS TO SUIT CASING 'A'

USE LEVEL

16"

VERTICAL FURRING MUST COME UNDER PANEL BUTT JOINTS IF USED AT ALL

SHIM IF NEEDED

16"

FURRING BEHIND BASE MOULD CAN BE SINGLE OR DOUBLE

INSIDE CORNERS AIM NAIL IN ON A SLANT TO ENTER STUDS

I" X 2" FURRING RUNS HORIZONTAL, AND IS NAILED TO EACH HIDDEN WALL STUD

OLD WALL

NEW

FURRING

REMOVE CASING

ADD CASING

FILLER STRIP

DOORWAY

ACCLIMATE PANELING

Since wood products do breath, it's a good idea on jobs like this to store the new wall covering in the room for a day or two before application. This gives it a chance to get accustomed to its new home. Also, before nailing, stand panels against the wall as you think they should be placed. This will give you a chance to change your mind for better grain-pattern effect *before* you start driving nails.

PANELING MASONRY WALLS

A typical application not covered by the above would be a masonry wall where you could not use ordinary fasteners. In such a case, furring could be attached with an adhesive, with special, hardened nails, with glued on masonry anchors or even toggle bolts. When the wall, or part of it, is below grade, it's a good idea to waterproof it before covering. This can be accomplished with a waterproofing cement paint.

If the walls are very rough or if you wish to avoid application directly to masonry, it would not be too much of a chore to build out a new wall for the paneling. This would be constructed like any rough framing with one exception. It will not be bearing wall so lighter materials are permissible. A 2×3, even 2×2 stock may be used. Attach top plate to the floor. If the floor is concrete, use case hardened nails to do the job.

Even when you build out a new wall this way, it's a good idea to waterproof the existing one.

Suspended redwood grid ceiling hangs from wires nailed to joists. You can hang it any-where from 3 to 24 inches below existing work. Main runners go at right angles to joists.

ACOUSTICAL CEILINGS

Sound-proofing tiles can be glued on, stapled to strips or suspended

Two methods of installation have be-come standards for homeowners. One is the stapling of flanged tiles to fur-ring strips—the other is to suspend a gridwork system that in turn supports the tiles. Either can be installed over an existing ceiling or directly to open joists.

The suspended system is especially good when you want to reduce the height of a room or when existing ductwork, pipes, wires, etc., can't be hidden by working directly to the joists. You can, of course, if it's appearance attracts you, use the system under any circumstances. The only consideration is that you must drop the new ceiling at least three inches.

MAKE SCALED DRAWING

For either method, your first step should be to make a scaled bird's-eye view of the room as shown in the draw-ing. This will tell you about tile place-ment and also provide a means for judg-ing material requirements. As you can see, the important consideration for tile placement is how you end up at borders. This last tile line, at all four walls, should not be less than ½ a tile width. You can, if you wish, test your layout by placing a line of tiles, face up, on the floor. Adjust the line one way or the other until the border tiles are equal. The distance from

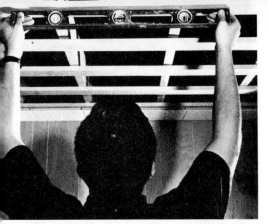

Check furring strips with a level, whether you are applying them against open joists or over an existing ceiling. Like hanging grid, they also run right angles to joists.

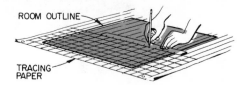

ROOM OUTLINE

TRACING PAPER

RULE 1" SQUARES ON PAPER FOR 12" TILES - ARRANGE PAPER OVER ROOM OUTLINE, SHIFTING TO DETERMINE HOW TO PLACE TILES, HOW MANY YOU MUST BUY -

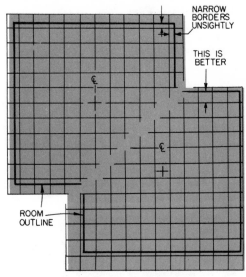

NARROW BORDERS UNSIGHTLY

THIS IS BETTER

ROOM OUTLINE

SHIFTING PAPER SHOWS TILES COME OUT BETTER WITH WIDER BORDER—THIS ALSO SHOWS YOU WHERE TO NAIL UP FURRING

Use thin shims where necessary to get the furring level. Drive two nails per joint crossing. Make sure joints fall on joists. Where you do have a joint, use four nails.

the joint of the border tile to the wall equals the distance from the wall to the center of the first furring strip—or to the center of the first grid runner.

FURRING STRIPS

Furring strips are dry softwood measuring about 1×3. They are nailed up at right angles to the joists, spaced center-to-center a distance equal to a tile width. The only exception, as we mentioned, is the strip closest to the walls. Use common or box nails long enough to penetrate the joists about 1½". Use two nails per joint crossing. Furring strips are also used at the ceiling line along all four walls. Work with a level when you set them; use thin strips of wood as shims, when necessary, to bulk out any low spots.

Start installation of the tiles in a corner, cutting border tiles to fit. Cutting can be done with a sharp knife or on a table saw if you own a special plywood cutting blade. This type of blade will do a good job on acoustical tile.

Use as many staples as instructions that come with the tiles tell you to. This can vary from product to product and also depends on tile size. Handle tiles carefully—they damage easily.

Start installation of tiles from a corner. Joint against wall is important unless you plan to cover with molding. Buy or rent a stapling gun, it will make work go faster.

Tile may be cut with a knife, but be sure it's a sharp one. Make a light cut to sever surface fibers, then make repeat cuts. Use plywood-cutting blade on table saw.

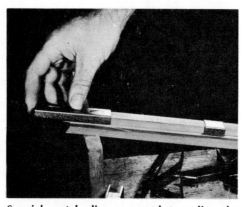

Special metal clips are used to splice the main runners. Other runners have pierced tabs to take free end of wire that is nailed to joist. Ceilings are light in weight.

A line stretched from wall to wall tells you where to bend or cut wire. When joists are covered and wire can't be nailed as shown, heavy screw-eyes can be utilized.

Tiles slip through openings created by the gridwork and are set on ledges in the grid pieces. On open wood gridwork, tiles are installed as short pieces are put up.

A finished, suspended ceiling looks like this. Gridwork is metal. Special straps which come with this Simpson system can be used. Notches allow entire grid to go up.

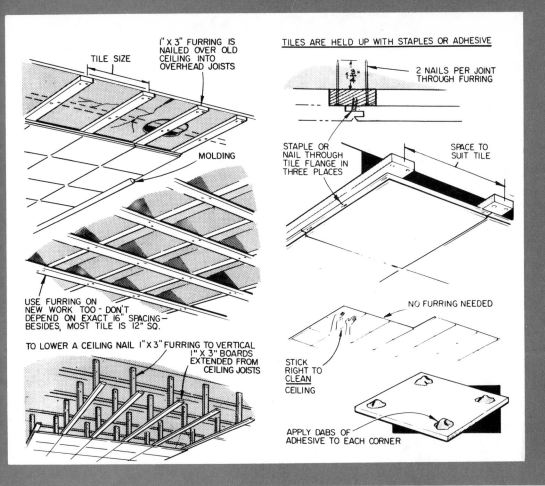

TILE SIZE

1" X 3" FURRING IS NAILED OVER OLD CEILING INTO OVERHEAD JOISTS

MOLDING

USE FURRING ON NEW WORK TOO - DON'T DEPEND ON EXACT 16" SPACING— BESIDES, MOST TILE IS 12" SQ.

TO LOWER A CEILING NAIL 1" X 3" FURRING TO VERTICAL 1" X 3" BOARDS EXTENDED FROM CEILING JOISTS

TILES ARE HELD UP WITH STAPLES OR ADHESIVE

2 NAILS PER JOINT THROUGH FURRING

STAPLE OR NAIL THROUGH TILE FLANGE IN THREE PLACES

SPACE TO SUIT TILE

NO FURRING NEEDED

STICK RIGHT TO CLEAN CEILING

APPLY DABS OF ADHESIVE TO EACH CORNER

SUSPENDED CEILINGS

Best way to start a suspended ceiling is to use a level to mark the new ceiling line on all four walls. Nail up the wall molding (which is part of the grid system) to follow the line. This gives you points from which you can do constant checking to be sure of alignment of other parts.

Most systems of this type hang from wire which is nailed to the side of the joist. How high up on the joist you go isn't important, but where you cut the wire is. To establish this cut line accurately, you stretch a string across the room using the wall molding as a guide. With some products the wire has to be bent at this point instead of being cut.

When you begin to put up the gridwork pieces, start with a center main-runner and work right across the room. End pieces are cut to fit snugly against the wall molding. Follow this procedure until all main runners are up. The short cross-runners come cut to length so all you do is fit them between the main runners. Again, start at the room center and run one line from wall to wall. To get the line straight, stretch a temporary guide line or work with a square. The end pieces are cut to fit and should be snug enough so the whole line is braced between walls. Now you can start putting tiles up as you continue to add lines of cross-runners.

WOOD JOINTS

Choose the easiest joint to make that will do the job adequately

What we've tried to do here is demonstrate a variety of joints for specific applications. In many instances there are more than several joints that may be used. So how do you make a choice? Well, to make a quick point, what tools do you own?

The mortise for a mortise-tenon joint is quite easy to do if you own a drill press and the necessary accessories. A dowel joint, which to a considerable degree has the same strength characteristics as the motise-tenon, can be accomplished with a portable electric drill or even a hand brace. Making dovetails with hand tools is an act of love while doing them with a portable router and a special jig is like so much production work. Fortunately, the variety of joints *is* available so doing a good job in this area is not a question of owning all the tools. A good substitute procedure for a dovetail (for example, on a drawer front) would be to rabbet the drawer-front for the drawer-side and then reinforce with dowels—or—even with screws.

Indoor furniture projects, especially when made with fancy woods, require more concern with appearance of joints than utility furniture. All joints should be strong, however.

Good-looking kitchen cabinetry, but if you check similar furniture in your kitchen you will discover that the bulk of the joints used are butts. Clean cuts keep joints sharp.

EASIEST AND ADEQUATE

Many times the right joint to use is not a question of equipment at all. If you see that all the joints are simple butts; something you can accomplish quite handsomely by working carefully with a handsaw. So another factor in selection is choosing the easiest-to-do that is adequate for the job. This makes more sense today than it ever did because of modern adhesives that often hold when adjacent parts break away.

As far as appearance is concerned, most joints show up as a single line between parts anyway. This is as true of a motise-tenon or a dowel joint as it is of a butt. What's inside the line (between the two members) has more to do with strength than appearance.

A good example of a joint used to hide a wood area is the miter. Its purpose is to join two pieces so that end grain is hidden, but its not an especially strong joint. If you compare it to a butt you find that it has a little more glue area because of the

This darkroom paper safe was built more for function than for looks, hence exposed rabbets, dadoes and plywood edges. This does not mean that it is not well built.

Many joints, like this cutaway of through mortise, can be reinforced with wedges. A wedge will force tenon tightly in cavity. Slot for wedge is previously cut in tenon.

Difference between a mortise and a simple hole is shown here. Mortising bit, which is a drill-press accessory, turns in a special square chisel to remove corners.

Miter joints can be made much stronger by using method shown—or by similar treatment. It may look complicated, but all it adds up to are simple, careful cuts on saw.

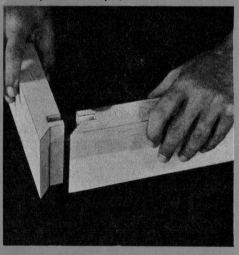

Dovetail joints are quite easy to do with a portable router and a special fixture. This is a great joint to use on dresser fronts, but time consuming to do by hand.

angle cut and this gives it a slight edge. It's a great joint for plywood since it conceals the structure of the material but making it really strong requires a spline or similar reinforcement.

PLYWOOD EDGES

Plywood edges always require more care because the plies are not attractive but there are many ways to hide them—

LEG TO RAIL JOINTS

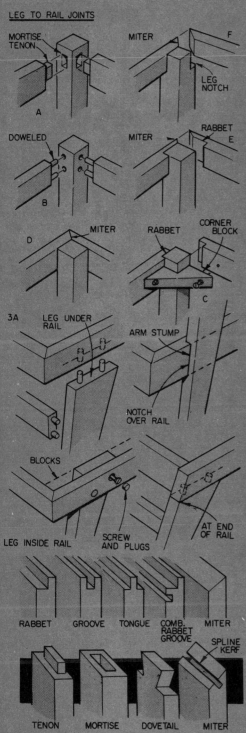

MORTISE
TENON

A

MITER

LEG
NOTCH

F

DOWELED

B

MITER

RABBET

E

D

MITER

RABBET

CORNER
BLOCK

C

3A

LEG UNDER
RAIL

ARM STUMP

NOTCH
OVER RAIL

BLOCKS

AT END
OF RAIL

LEG INSIDE RAIL

SCREW
AND PLUGS

RABBET   GROOVE   TONGUE   COMB.
RABBET
GROOVE   MITER

SPLINE
KERF

TENON   MORTISE   DOVETAIL   MITER

ALL EXCEPT MORTISE CAN BE CUT WITH
SAW BLADE, BUT DADO BLADE IS BEST

CASE JOINTS WHERE TOP PROJECTS

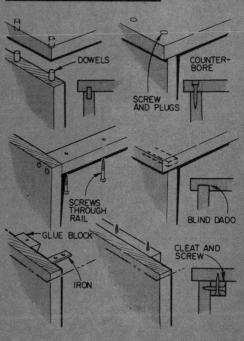

DOWELS

COUNTER-
BORE

SCREW
AND PLUGS

SCREWS
THROUGH
RAIL

BLIND DADO

GLUE BLOCK

CLEAT AND
SCREW

IRON

JOINING TWO RAILS AT INTERSECTION

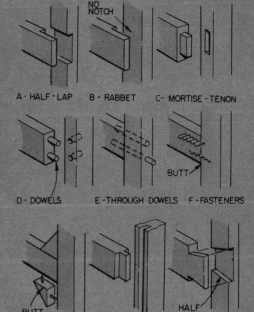

NO
NOTCH

A - HALF - LAP   B - RABBET   C- MORTISE - TENON

BUTT

D - DOWELS   E - THROUGH DOWELS   F - FASTENERS

BUTT

HALF
LAP

G - GLUE BLOCKS   H - TONGUE - GROOVE   I - DOVETAIL

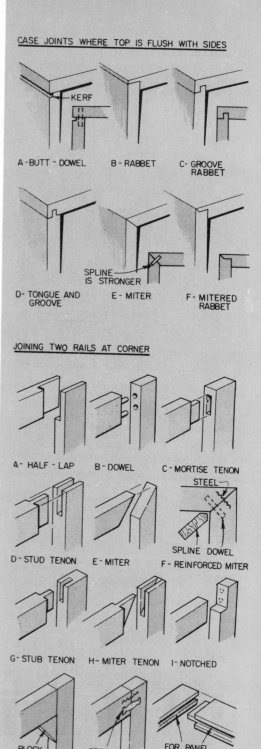

KERF

A - BUTT - DOWEL   B - RABBET   C - GROOVE RABBET

SPLINE IS STRONGER

D - TONGUE AND GROOVE   E - MITER   F - MITERED RABBET

The chunkiness of this table is achieved by treating each square leg as a box and proceed with mitered edges, spline joints. Can be in painted or formica-covered ply.

JOINING TWO RAILS AT CORNER

A - HALF - LAP   B - DOWEL   C - MORTISE TENON

STEEL

SPLINE DOWEL

D - STUD TENON   E - MITER   F - REINFORCED MITER

G - STUB TENON   H - MITER TENON   I - NOTCHED

BLOCK   STEEL   FOR PANEL INSERTION

J - SIMPLE BUTT   L - BUTT - REINF.   K - TONGUE - GROOVE

Fine furniture like this cabinet designed by Leo Jiranek, A.I.D. and built by Bassett Industries, incorporates a myriad of joint types, all designed for a purpose.

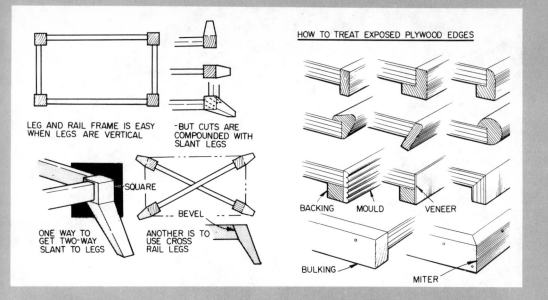

LEG AND RAIL FRAME IS EASY WHEN LEGS ARE VERTICAL

-BUT CUTS ARE COMPOUNDED WITH SLANT LEGS

SQUARE

ONE WAY TO GET TWO-WAY SLANT TO LEGS

BEVEL

ANOTHER IS TO USE CROSS RAIL LEGS

HOW TO TREAT EXPOSED PLYWOOD EDGES

BACKING   MOULD   VENEER

BULKING

MITER

running from commercial banding that you simply press on to gluing on molding that you can make or buy. The sketch shows many ways to accomplish this. Bulking the edge is just a matter of using material that is heavier than the plywood thickness. This not only hides the edges but gives the panel more visual substance when you use it, for example, as a table top. A good, square edge on the plywood and the edging material is essential to keep the joint line as fine as possible.

HOW MUCH ABUSE

The kind of abuse the joint will take is another factor in selection. The strain on a drawer is mostly where the drawer sides meet the drawer-front. The more weight in the drawer the greater the strain at that point when you pull the drawer out. So—this area in a shop drawer that will hold heavy tools is more critical than it would be in a drawer designed to hold stamps or pencils.

Table or chair legs that angle out are more likely to separate where they meet the rails than legs that are vertical. Extra strength can be provided not only in the leg-to-rail joint but in the use of stretchers between the legs. A similar example is in a sawhorse. Legs are angled out to give the project stability but weight puts considerable strain where the legs meet the horizontal rail. To counteract this effect, install stretcher across the legs.

ACCURACY IS IMPORTANT

Accuracy is always important—for strength as well as looks. A dowel in a hole too large for it is not going to hold as well as one that fits snugly. On the other hand, force-fitting a dowel can do more harm than good. Dowel pegs should be a slip-fit in the holes and should be chamfered at each end and grooved (or spiraled) along the length so that excess glue can escape. If this isn't done and the fit is a tight one, you can exert enough pressure under clamps to force the glue out through pores in the wood—even to crack the wood.

Don't rely on clamp pressure to bring parts together or wood putty to fill a gap. Wood dough products have a place in the shop but they are not intended to hide careless work or to add strength to a poor fit.

Projects you make yourself can be creative in depth. The top of this coffee table was a cut-out from a discarded solid-core door purchased for $1.00 from a local door maker.

FURNITURE BUILDING

Design for beauty, function and harmony of materials and hardware

Designing a piece of furniture from scratch involves no more than some pre-planning concerning its function. Once, when Mrs. D. wanted a chair exactly right for her, I had her get down on the floor on her side in a sitting posi-tion and traced her profile on a large piece of wrapping paper. Maybe she felt silly at the time but the chair was very successful and about as "custom" a design as you can get.

The basic design, regardless of what

you make, has to be for function. A chair may be beautiful but if you bang your knees when sitting in it at the table, the chair (or maybe the table) is wrong. What this does is bring successful designing within the scope of everyone since it's just a matter of simple logic. I often use a desk project as a means of illustrating the idea because its a good example of designing through function.

PROPER DIMENSIONS

If you sat in a chair and experimented with bricks and a plank to discover a good height for a comfortable writing surface you would find it to be not much more or less than 29''. So you see the height is not arbitrary. Stay in the chair and imagine a desk around you. Reach forward as far as you can and this would establish a maximum depth. Stretch your arms and this will tell you the maximum length. Draw these three dimensions on a piece of paper and you have a form in which you design the desk. Kneehole

PROJECTS AND DIMENSIONS (all dimension in inches)			
NAME	A	B	C
kitchen stool	24	14	14
coffee table	20	36	16
tea cart	25	30	20
step stool	32	23	16
bookcase	36	48	12
magazine rack	18	14	10
dressing table	26	40	18
newspaper rack	18	26	10
telephone stand	28	20	16
telephone shelf	12	12	11
bedside stand	26	18	14
table-top chest	6	14	10
chest	18	36	18
spice cabinet	12	16	5
cutting blocks	3	12	16
spice shelf	12	16	4
5-board stool	7	16	6½
5-board bench	17	36	14
room divider	48	50	14
step table	25	20	26
night stand	26	18	18
desk caddy	16	32	13
shadow box	16	38	5
planter	32	32	10
folding screen	68-72	16-18	

A-B-C basic dimensions in above chart are applicable to drawing at left and others through pg. 105. Other dimensions optional.

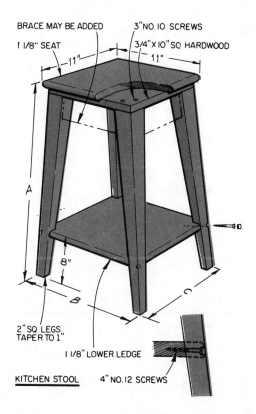

BRACE MAY BE ADDED   3"NO.10 SCREWS
1 1/8" SEAT   3/4"X 10" SQ HARDWOOD
—11"—   — 11"—
A
8"
B
C
2" SQ. LEGS, TAPER TO 1"
1 1/8" LOWER LEDGE
KITCHEN STOOL   4" NO.12 SCREWS

height? Measure from the floor to the top of your knees and that gives you the minimum. Subtract this from the total height and you know how much room you have for a pencil drawer. By measuring across your lap, you get the minimum dimension for the width of the kneehole opening.

Length and depth can change depending on what the desk will be used for. A little thing for the kitchen doesn't have to be sized like one for the office or den, but the height and the kneehole

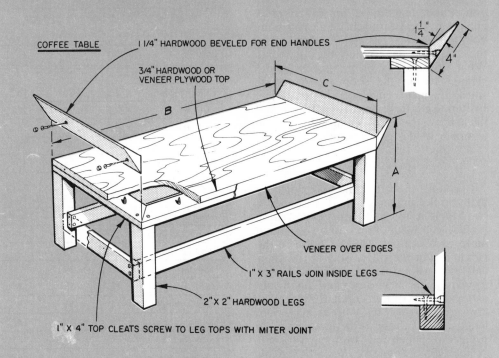

COFFEE TABLE

1 1/4" HARDWOOD BEVELED FOR END HANDLES

3/4" HARDWOOD OR VENEER PLYWOOD TOP

B

C

A

1 1/4"

4"

VENEER OVER EDGES

1" X 3" RAILS JOIN INSIDE LEGS

2" X 2" HARDWOOD LEGS

1" X 4" TOP CLEATS SCREW TO LEG TOPS WITH MITER JOINT

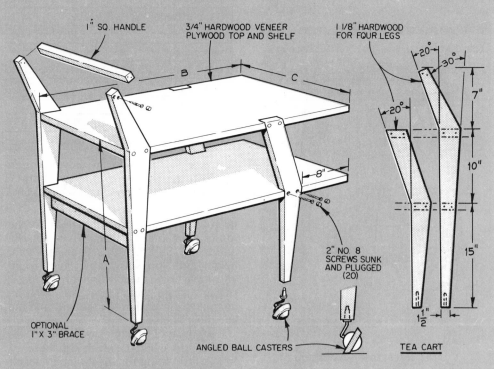

1" SQ. HANDLE

3/4" HARDWOOD VENEER PLYWOOD TOP AND SHELF

1 1/8" HARDWOOD FOR FOUR LEGS

B

C

20°

30°

20°

7"

10"

15"

8"

2" NO. 8 SCREWS SUNK AND PLUGGED (20)

A

OPTIONAL 1" X 3" BRACE

ANGLED BALL CASTERS

1 1/2"

TEA CART

A-B-C dimensions on drawings across these pages relate to the basic size chart on page 97.

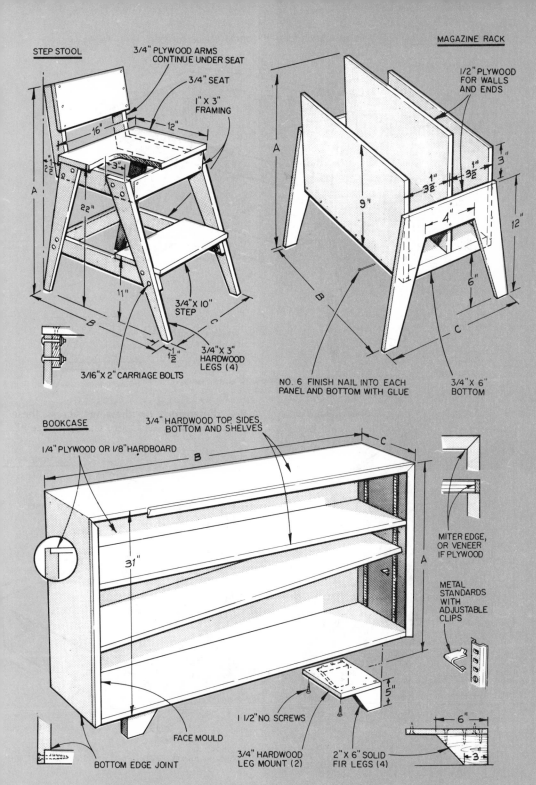

STEP STOOL

3/4" PLYWOOD ARMS
CONTINUE UNDER SEAT

3/4" SEAT

1" X 3"
FRAMING

16"

12"

3"

1
2
2

A

22"

11"

B

C

1
2

3/4"X 10"
STEP

3/4"X 3"
HARDWOOD
LEGS (4)

3/16"X 2" CARRIAGE BOLTS

MAGAZINE RACK

1/2" PLYWOOD
FOR WALLS
AND ENDS

A

A

9"

1
3 2

1"
3 2

3"

12"

4"

6"

B

C

NO. 6 FINISH NAIL INTO EACH
PANEL AND BOTTOM WITH GLUE

3/4" X 6"
BOTTOM

BOOKCASE

3/4" HARDWOOD TOP, SIDES,
BOTTOM AND SHELVES

1/4" PLYWOOD OR 1/8" HARDBOARD

B

C

A

31"

MITER EDGE,
OR VENEER
IF PLYWOOD

METAL
STANDARDS
WITH
ADJUSTABLE
CLIPS

5"

FACE MOULD

1 1/2" NO. SCREWS

3/4" HARDWOOD
LEG MOUNT (2)

2" X 6" SOLID
FIR LEGS (4)

6"

3

BOTTOM EDGE JOINT

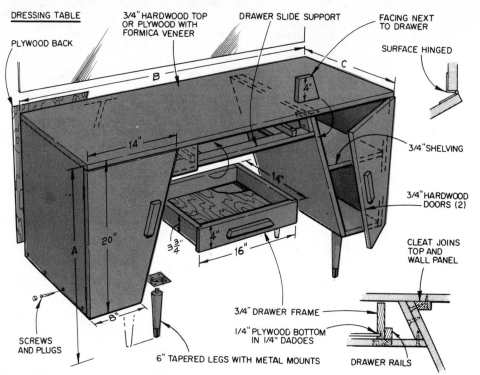

PLYWOOD BACK

3/4" HARDWOOD TOP OR PLYWOOD WITH FORMICA VENEER

DRAWER SLIDE SUPPORT

FACING NEXT TO DRAWER

SURFACE HINGED

B

C

4"

14"

14"

3/4" SHELVING

20"

3¾"

4"

16"

3/4" HARDWOOD DOORS (2)

CLEAT JOINS TOP AND WALL PANEL

A

3/4" DRAWER FRAME

1/4" PLYWOOD BOTTOM IN 1/4" DADOES

SCREWS AND PLUGS

8"

6" TAPERED LEGS WITH METAL MOUNTS

DRAWER RAILS

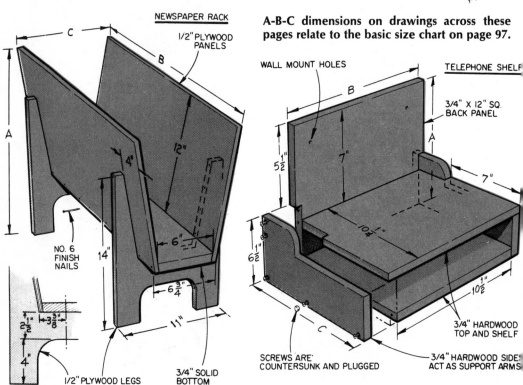

NEWSPAPER RACK

1/2" PLYWOOD PANELS

A-B-C dimensions on drawings across these pages relate to the basic size chart on page 97.

C

B

WALL MOUNT HOLES

TELEPHONE SHELF

B

3/4" X 12" SQ. BACK PANEL

A

4"

12"

A

5½"

7"

7"

6"

10¼"

NO. 6 FINISH NAILS

14"

6¾"

11"

6½"

1"

10½"

2½" 3/8"

4"

1/2" PLYWOOD LEGS

3/4" SOLID BOTTOM

SCREWS ARE COUNTERSUNK AND PLUGGED

3/4" HARDWOOD TOP AND SHELF

3/4" HARDWOOD SIDE ACT AS SUPPORT ARMS

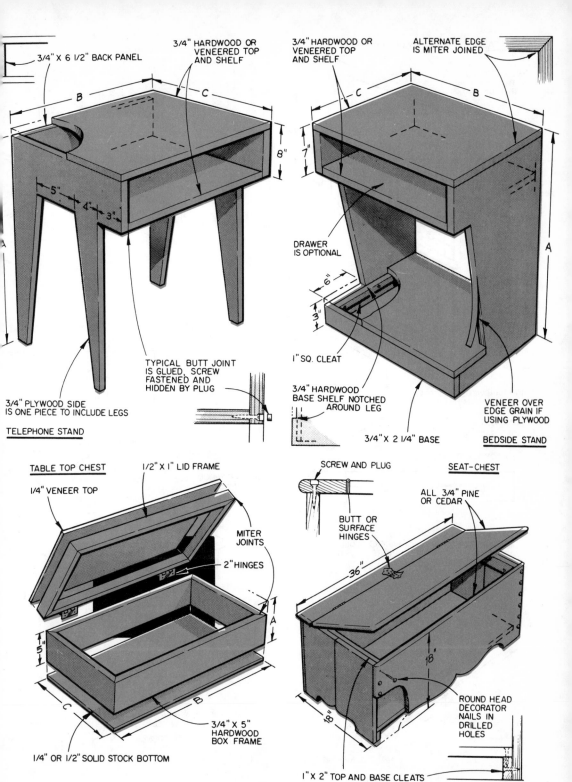

3/4" X 6 1/2" BACK PANEL

3/4" HARDWOOD OR VENEERED TOP AND SHELF

3/4" HARDWOOD OR VENEERED TOP AND SHELF

ALTERNATE EDGE IS MITER JOINED

B

C

8"

TYPICAL BUTT JOINT IS GLUED, SCREW FASTENED AND HIDDEN BY PLUG

3/4" PLYWOOD SIDE IS ONE PIECE TO INCLUDE LEGS

TELEPHONE STAND

5"

4"

3"

C

B

7"

A

DRAWER IS OPTIONAL

1" SQ. CLEAT

3/4" HARDWOOD BASE SHELF NOTCHED AROUND LEG

6"

3"

3/4" X 2 1/4" BASE

VENEER OVER EDGE GRAIN IF USING PLYWOOD

BEDSIDE STAND

TABLE TOP CHEST

1/2" X 1" LID FRAME

1/4" VENEER TOP

MITER JOINTS

2" HINGES

A

5"

C

B

3/4" X 5" HARDWOOD BOX FRAME

1/4" OR 1/2" SOLID STOCK BOTTOM

SCREW AND PLUG

SEAT-CHEST

BUTT OR SURFACE HINGES

ALL 3/4" PINE OR CEDAR

36"

18"

18"

ROUND HEAD DECORATOR NAILS IN DRILLED HOLES

1" X 2" TOP AND BASE CLEATS

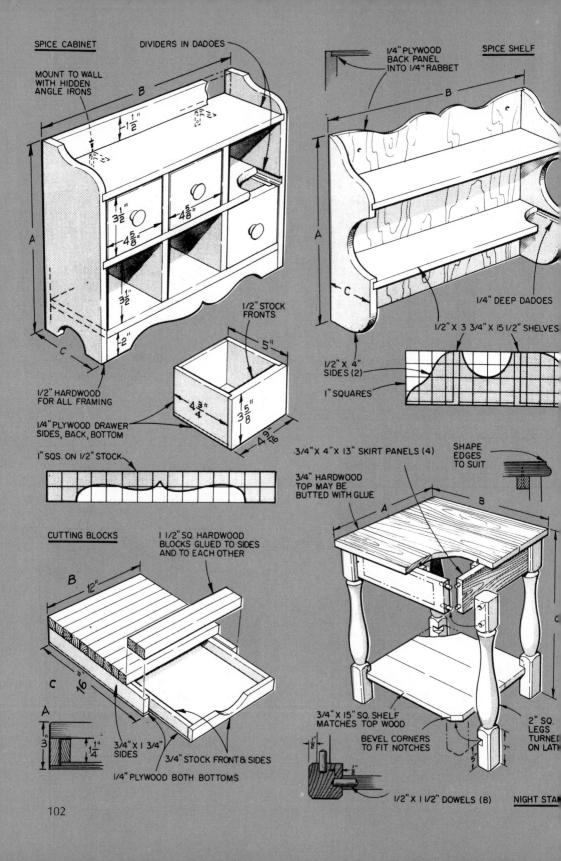

SPICE CABINET

DIVIDERS IN DADOES

MOUNT TO WALL
WITH HIDDEN
ANGLE IRONS

B

$1\frac{1}{2}$"

$3\frac{1}{2}$"

$4\frac{5}{8}$"

$4\frac{5}{8}$"

$3\frac{1}{2}$"

A

C

2"

1/2" HARDWOOD
FOR ALL FRAMING

1/4" PLYWOOD DRAWER
SIDES, BACK, BOTTOM

1" SQS. ON 1/2" STOCK

1/2" STOCK
FRONTS

5"

$4\frac{3}{4}$"

$3\frac{5}{8}$"

$4\frac{9}{16}$"

SPICE SHELF

1/4" PLYWOOD
BACK PANEL
INTO 1/4" RABBET

B

A

C

1/4" DEEP DADOES

1/2" X 3 3/4" X 15 1/2" SHELVES

1/2" X 4"
SIDES (2)

1" SQUARES

3/4"X 4"X 13" SKIRT PANELS (4)

3/4" HARDWOOD
TOP MAY BE
BUTTED WITH GLUE

SHAPE
EDGES
TO SUIT

CUTTING BLOCKS

1 1/2" SQ. HARDWOOD
BLOCKS GLUED TO SIDES
AND TO EACH OTHER

B

12"

C

16"

A

3"

$1\frac{1}{4}$"

3/4" X 1 3/4"
SIDES

3/4" STOCK FRONT & SIDES

1/4" PLYWOOD BOTH BOTTOMS

A

B

3/4" X 15" SQ. SHELF
MATCHES TOP WOOD

BEVEL CORNERS
TO FIT NOTCHES

2" SQ.
LEGS
TURNED
ON LATHE

$\frac{5}{?}$"

1/2" X 1 1/2" DOWELS (8)

NIGHT STA

102

A-B-C dimensions on drawings across these pages relate to basic size chart on page 97.

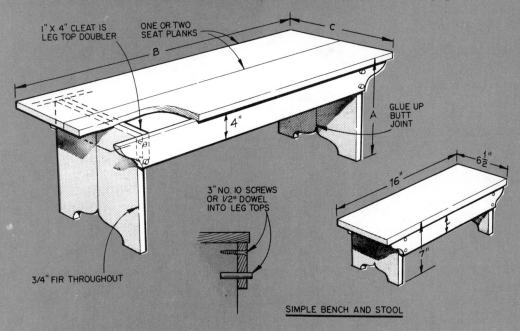

1" X 4" CLEAT IS LEG TOP DOUBLER

ONE OR TWO SEAT PLANKS

B

C

4"

A

GLUE UP BUTT JOINT

3" NO. 10 SCREWS OR 1/2" DOWEL INTO LEG TOPS

3/4" FIR THROUGHOUT

6 1/2"

16"

7"

SIMPLE BENCH AND STOOL

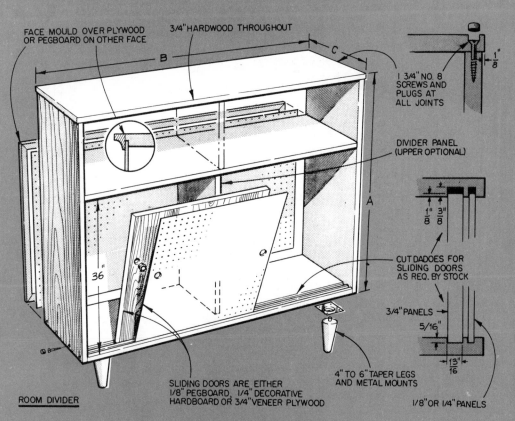

FACE MOULD OVER PLYWOOD OR PEGBOARD ON OTHER FACE

3/4" HARDWOOD THROUGHOUT

B

C

1 3/4" NO. 8 SCREWS AND PLUGS AT ALL JOINTS

1/8"

DIVIDER PANEL (UPPER OPTIONAL)

A

1/8" 3/8"

36

CUT DADOES FOR SLIDING DOORS AS REQ. BY STOCK

3/4" PANELS

5/16"

4" TO 6" TAPER LEGS AND METAL MOUNTS

13/16"

ROOM DIVIDER

SLIDING DOORS ARE EITHER 1/8" PEGBOARD, 1/4" DECORATIVE HARDBOARD OR 3/4" VENEER PLYWOOD

1/8" OR 1/4" PANELS

103

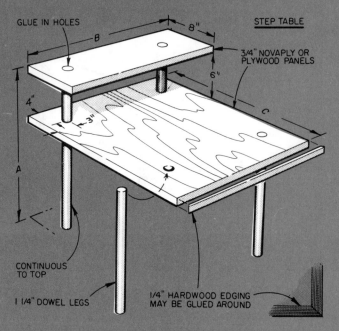

STEP TABLE

GLUE IN HOLES

B

8"

3/4" NOVAPLY OR
PLYWOOD PANELS

6"

4"

3"

C

A

CONTINUOUS
TO TOP

1 1/4" DOWEL LEGS

1/4" HARDWOOD EDGING
MAY BE GLUED AROUND

A-B-C dimensions on drawings across these pages relate to basic size chart on page 97.

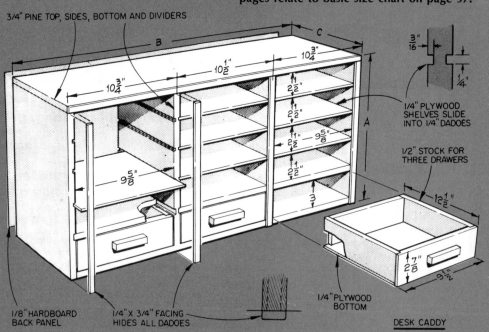

3/4" PINE TOP, SIDES, BOTTOM AND DIVIDERS

B

C

3/16"

1/4"

10 3/4"

10 3/4"

10 1/2"

2 1/2"

2 1/2"

1/4" PLYWOOD
SHELVES SLIDE
INTO 1/4" DADOES

A

9 5/8"

2 1/2"

9 5/8"

1/2" STOCK FOR
THREE DRAWERS

9 5/8"

2 1/2"

3

12 1/2"

2 7/8"

9 1/2"

1/4" PLYWOOD
BOTTOM

1/8" HARDBOARD
BACK PANEL

1/4" x 3/4" FACING
HIDES ALL DADOES

DESK CADDY

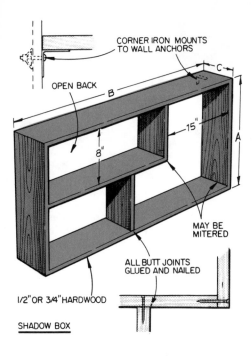

CORNER IRON MOUNTS TO WALL ANCHORS

OPEN BACK

B

15"

8"

A

C

MAY BE MITERED

ALL BUTT JOINTS GLUED AND NAILED

1/2" OR 3/4" HARDWOOD

SHADOW BOX

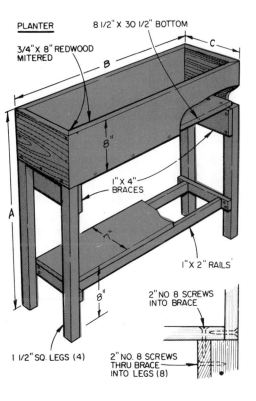

PLANTER

8 1/2" X 30 1/2" BOTTOM

3/4" X 8" REDWOOD MITERED

B

C

8"

I" X 4" BRACES

A

I" X 2" RAILS

8"

2" NO 8 SCREWS INTO BRACE

1 1/2" SQ. LEGS (4)

2" NO. 8 SCREWS THRU BRACE INTO LEGS (8)

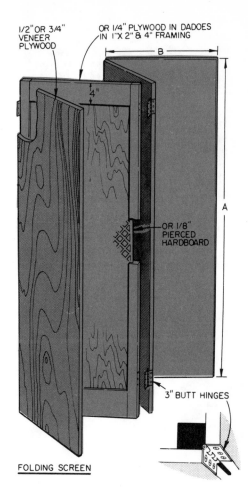

1/2" OR 3/4" VENEER PLYWOOD

OR 1/4" PLYWOOD IN DADOES IN 1"X 2" & 4" FRAMING

B

4"

A

OR 1/8" PIERCED HARDBOARD

3" BUTT HINGES

FOLDING SCREEN

room shouldn't change unless you are custom-designing one for a child.

PROPER DESIGN

How do you design for storage in the desk? Well, what must the desk hold? A pad and pencil for kitchen use or for near a telephone? Or will it be a place to do typing, handle household accounts, file documents, store pencils and pens, the stapling machine, etc. The answers to these questions can lead anywhere from a simple slab on readymade legs to a conventional type unit.

Designing *strictly* for function can be a style in itself but the choice of materials, the kind of finish, use of moldings or hand-carvings, choice of hardware,

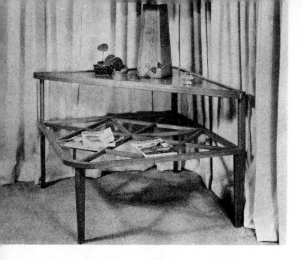

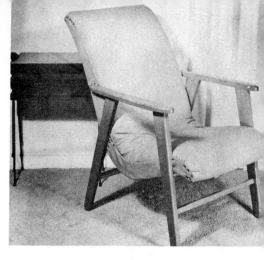

This project incorporates a fancy plywood and matching lumber. Numerous joints, some at odd angles, make this difficult job. Available time can influence the design.

Chairs, especially upholstered ones, are more difficult projects. Don't tackle jobs like this unless you own bandsaw, jigsaw or saber saw. Hardy ones can use handsaw.

"Sculptured" joints are matter of smoothing parts together to point where they might resemble one piece. Any joint can be sculptured, but it usually means much hand work.

Bevel cuts on table saw produce segments that can be joined for round objects. Set number of segments, divide into 360 degrees, then set table-saw tilt to half sum.

even the joints you use bring you into the area where you consider looks or compatability with an established decor. Using moldings on one can make two identical projects look completely different yet the design-for-function aspect would not be affected.

FURNITURE MATERIALS

The material you choose plays a top role. Knotty pine and maple go along with exposed dowel pegs and rounded corners—early American and colonial pieces. Not so with aluminum angle and

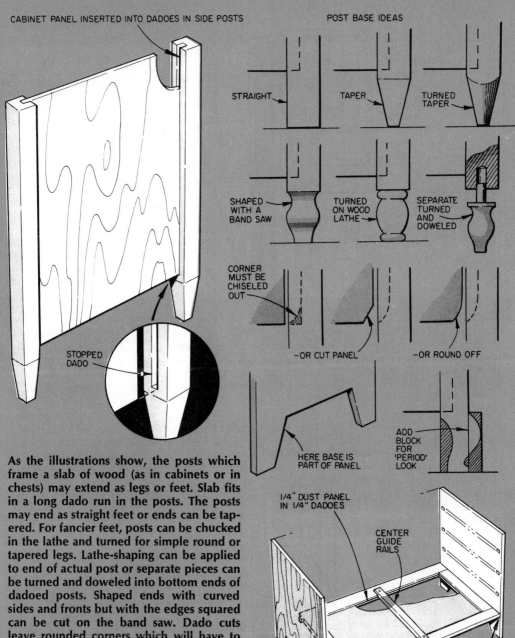

CABINET PANEL INSERTED INTO DADOES IN SIDE POSTS

POST BASE IDEAS

STRAIGHT

TAPER

TURNED TAPER

SHAPED WITH A BAND SAW

TURNED ON WOOD LATHE

SEPARATE TURNED AND DOWELED

CORNER MUST BE CHISELED OUT

—OR CUT PANEL

—OR ROUND OFF

STOPPED DADO

HERE BASE IS PART OF PANEL

ADD BLOCK FOR 'PERIOD' LOOK

1/4" DUST PANEL IN 1/4" DADOES

CENTER GUIDE RAILS

1" X 3" DRAWER SUPPORT FRAME

TENONED ENDS GLUED INTO MORTISES

BASIC FRAME FOR CASE CONSTRUCTION

As the illustrations show, the posts which frame a slab of wood (as in cabinets or in chests) may extend as legs or feet. Slab fits in a long dado run in the posts. The posts may end as straight feet or ends can be tapered. For fancier feet, posts can be chucked in the lathe and turned for simple round or tapered legs. Lathe-shaping can be applied to end of actual post or separate pieces can be turned and doweled into bottom ends of dadoed posts. Shaped ends with curved sides and fronts but with the edges squared can be cut on the band saw. Dado cuts leave rounded corners which will have to be chiseled out—or square edges of panel can be rounded off to fit dadoes. Basic frame case construction without any corner posts permits slanted cut-outs on bottom edges to make self-legs as shown. Period-piece design can be achieved by a blocking out of the legs (shaded areas) to bulk out and round off design. Drawer support frames can have tenons on corners to fit into matching mortises on side panels as shown in drawings.

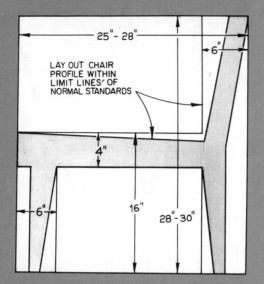

Plan a chair within a form established by standard chair dimensions as shown above. You don't have to stick to these, exactly, but they are a good place to begin. Now make rough sketches for flair ideas. The application of ideas must now fit limits of standards.

Draw basic outlines of chair parts within the limit lines. Don't expect your first attempts to be completely successful. The simpler the design the easier it will be to translate it into actual construction. Straight lines, of course, are the easiest. Note seat, arm lines parallel.

You can work with curves, but even though they may be easy to cut out of the actual plywood, for example, the assembly regarding joints could get a bit complicated for first attempts. Note the jointing and the placement of dowels in the example shown here.

Stick to simple design at first, or work with straight lines for basic profile and go on from there. The straight-line chair shown below can be made in solid hardwood or even out of square metal "tubes" as is shown in the drawing. Assembly of the parts is easy.

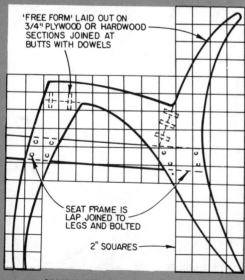

PLYWOOD FREE FORM CONTOURS

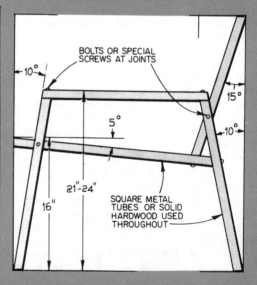

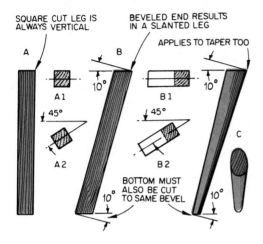

SQUARE CUT LEG IS ALWAYS VERTICAL

BEVELED END RESULTS IN A SLANTED LEG

APPLIES TO TAPER TOO

A

B

A1

10°

B1

10°

45°

45°

A2

B2

C

BOTTOM MUST ALSO BE CUT TO SAME BEVEL

10°

10°

A good example of case-goods construction is shown in the photo, above. Attention to sanding even before you begin to assemble all the parts is a recommended procedure.

After the design for your chair is ready for assembly, give some consideration to the problem of legs and their construction. The illustrations, above, show that when a leg is cut square at ends (A) it can't be anything but vertical regardless of whether it is set square as in (A-1) or at an angle as in (A-2). When a slant cut is made (B), the leg can slant at a simple angle (B-1) or at a compound angle, (B-2). This applies whether the leg is straight, tapered, square or round (C). Be sure top and bottom cuts are parallel.

teak. Where do you envision wrought iron and thick, rough-hewn pine boards? This doesn't mean you can't go off-trail. Many a decorator or conversation piece in a room doesn't fit when style alone is considered.

But on the whole, harmony is important; not only in terms of placement but when considering project details. Plastic, sliding doors wouldn't look good on an early american pine cabinet, not unless you were making a modern version of the piece. Heavy, wrought iron hinges don't seem right for sleek teak. On the practical side, you would not use a hand-size knob on a tiny stamp drawer anymore than you would install a fingertip pull on a big shop-drawer.

FURNITURE MODELS

You can go a long way before building

by doodling with a pencil—maybe even visiting furniture stores to get ideas—if you need them. Another thought is to make small cardboard models. This is an especially good idea when working with an expensive plywood since the cardboard components can be used on a scaled drawing of the plywood sheet to get best use of the material and also to plan grain direction.

This section includes 24 sketches of typical household projects together with basic dimensions. Armed with this and the ideas in the section on JOINTS should keep you workshopping for quite a while.

Good craftsmanship is not a mystery—its 90% *careful* work. Take the time to measure, to check, TWICE before you cut. Adopt this same pace throughout the project and you're bound to come up with nice things.

ABOUT BUILT-INS

Customize your house and get double service out of rooms, storage

Although the name has come to cover a multitude of projects, the true built-in becomes a permanent part of the house and, ideally, occupies space that could not be completely utilized otherwise. Quite often, existing walls, floor and ceiling are part of the project. For example, if you installed shelves and vertical dividers on one wall of a room, then faced them and added doors, you would have a true built-in and the unit would utilize existing surfaces as top, bottom and sides.

WHERE TO BUILD-IN?

Chances are you can find ideal locations in your home that afford the opportunity to do a good job with a built-in. Space under stairways, nooks made by a fire place that juts into the room, a passageway that has more width than it needs, maybe a jog in the wall caused by a closet in the next room. These sites are great because, exclusive of how you design the interior, they need no more than front framing, facing and doors to become built-ins.

ROOM DIVIDERS

Quite often a divider-type project—one used to define areas or to direct traffic—is called a built-in even though you design it so it can be moved. Perhaps the term has become more generally applicable because it can apply to items that you build in as well as to the project itself. Typical in this area would be hi-fi components, TV sets and the like.

WORK WITH A LEVEL

The fact that most times you do work with room-walls makes this facet of the construction necessary—achieve alignment of vertical and horizontal members by working with a level instead of a square or a ruler. Suppose you measured up from a floor at two points to mark a line for a shelf? The shelf would be parallel to the floor but not necessarily level. Better to mark up from the floor at *one* point, then use a level to carry the line across the wall. If you cut a board perfectly square, will it fit the corner made by two adjacent walls? Possibly—but

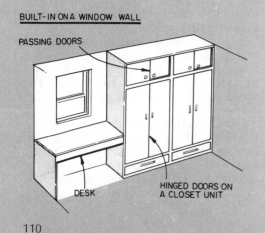

BUILT-IN ON A WINDOW WALL

PASSING DOORS

DESK

HINGED DOORS ON A CLOSET UNIT

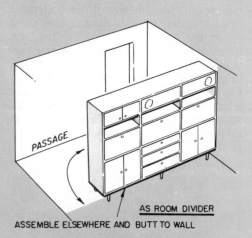

PASSAGE

AS ROOM DIVIDER

ASSEMBLE ELSEWHERE AND BUTT TO WALL

This built-in wall opens up to turn a roomy den into a useful dining room. Even the table folds up to become center panel of the unit. Cabinets hold flatware, other items.

it's something you should check before you cut.

Don't fret if you find walls that are not plumb—or more or less than 90° angles between walls or between wall and ceiling or wall and floor. You're lucky if you don't find these little discrepancies but they are "normal" in most homes. Be aware of them—*work with a level*—and you will have no problems.

BUILT-IN FRAMING

You can use "open" or "closed" framing. Picture the open framing as a skeleton structure that you then cover—the closed design as solid panels attached to each other with none or a minimum of support work. The sketches show examples of each. Making a choice between

the two is mostly a matter of project placement. By way of example—if you were building against the entire wall of a room, the skeleton framing would work out fine. If the project were centered on a wall; a unit with its own sides but utilizing the wall as its back, solid framing might work out better. Actually, either idea will work in any situation but basing a choice on the above factors will usually save time and/or material.

PRE-FABBING

Of course, it's also possible to pre-fab a built-in in the shop, then bring it in for permanent attachment in the house. This may very well be a good idea for a divider type unit that you wish to bring with you when you move. The project

111

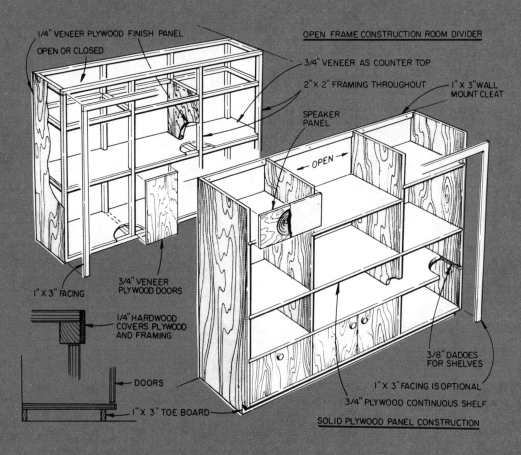

1/4" VENEER PLYWOOD FINISH PANEL
OPEN OR CLOSED

OPEN FRAME CONSTRUCTION ROOM DIVIDER

3/4" VENEER AS COUNTER TOP

2" X 2" FRAMING THROUGHOUT

1" X 3"WALL MOUNT CLEAT

SPEAKER PANEL

OPEN

1" X 3" FACING

3/4" VENEER PLYWOOD DOORS

1/4" HARDWOOD COVERS PLYWOOD AND FRAMING

DOORS

1" X 3" TOE BOARD

3/8" DADOES FOR SHELVES

1" X 3" FACING IS OPTIONAL

3/4" PLYWOOD CONTINUOUS SHELF

SOLID PLYWOOD PANEL CONSTRUCTION

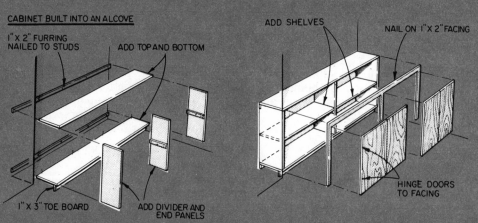

CABINET BUILT INTO AN ALCOVE

1" X 2" FURRING NAILED TO STUDS

ADD TOP AND BOTTOM

1" X 3" TOE BOARD

ADD DIVIDER AND END PANELS

ADD SHELVES

NAIL ON 1" X 2" FACING

HINGE DOORS TO FACING

First step here was furring strip nailed into studs through wall. This established top point to work to. In conventional room the ceiling would have been top of unit.

The two vertical pieces that outline the door opening were positioned with temporary cleats across front. Work between the verticals was framing for shelves, front.

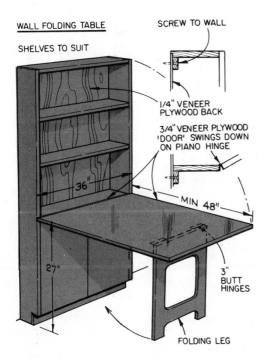

WALL FOLDING TABLE

SHELVES TO SUIT

SCREW TO WALL

1/4" VENEER PLYWOOD BACK

3/4" VENEER PLYWOOD 'DOOR' SWINGS DOWN ON PIANO HINGE

36"

MIN 48"

27"

3" BUTT HINGES

FOLDING LEG

SHELVES BUILT RIGHT INTO WALL NICHE

SAW AWAY PLASTER UP AGAINST STUDS

NAILS

BOX BUILT TO FIT FLUSH WITH WALL

SHELF TO SUIT

ADD FACE FRAMING

CUT

CAN BE PASS-THRU

IF A STUD IS CUT OUT A HEADER OF 2"X 4"S MUST BE ADDED

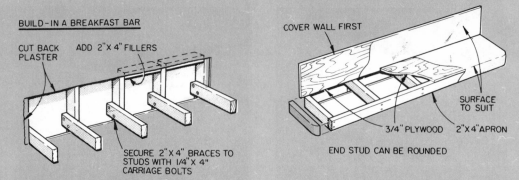

BUILD-IN A BREAKFAST BAR

CUT BACK PLASTER

ADD 2"X 4" FILLERS

SECURE 2"X 4" BRACES TO STUDS WITH 1/4"X 4" CARRIAGE BOLTS

COVER WALL FIRST

SURFACE TO SUIT

3/4" PLYWOOD

2"X 4"APRON

END STUD CAN BE ROUNDED

Large kitchen with free-standing wall for service island gains counter for snacks by adding the built-in shelf shown in drawing, above, and photo, right. Surface to suit.

Jutting fireplace wall provided ideal nook for built-in. In this case, area above base cabinet is more for looks than storage. Sketch shows how to accomplish jobs like this.

can be designed in modular units for stacking in position. Such an idea makes it possible, too, to "redesign" the unit when you wish, merely by changing position of the components. In such a case, picture each module as an independent piece but with sizes and shapes that will be compatible when they are stacked. This is not a difficult chore if you work on paper first, making a simple line drawing that indicates the unit as a whole. In fact, and we've done this often, after that initial drawing, make small cardboard models or even cut blocks of wood to represent the various sections. As far as design is concerned always plan the larger units for placement at the base. This will give the more visual stability.

CABINETS-IN-THE-WALLS

Typical of small, utility-type built-ins is the one in the sketch that shows how to install a cabinet *in* the wall. This is a rather simple job unless you want the project to be wide enough to make it necessary to remove part of a stud. In that case you do have to put in a "header" and a "sill". For a between-the-studs opening, locate one stud by tapping with a hammer on the wall. Pierce the wall with a saber saw or a keyhole saw at a point to the left or right of the stud location. Saw toward the stud until you hit it—then saw in the other direction

until you hit the next stud. Work with a level to mark the size of the opening you wish and then cut on the line to remove the covering. Follow the details in the sketch to finish the job.

CUSTOM-DESIGNED

Whatever or wherever, your built-in should be as custom-designed as possible. You may have to compromise *outside* so the unit will harmonize with its surroundings—but *inside* it's your baby and should be an outstanding example of space utilization. Why are you building the unit in the first place? If I guessed to provide more storage space or to provide a cabinet for something you want to build-in, I would almost certainly have to be right. So, as far as function is concerned it's just a question of designing around the items you are going to store. Arbitrary placement of fixed shelves for example, is never a good idea. To paint a picture, just visualize the average closet which is seldom more than a hole in the wall—then see it as a built-in with special places for the things it must contain. Your success with built-ins is based on nothing more than that.

The sections in this book on *furniture, wood joints, shelves* and *closets* contain much information that can be put to practical use when you are designing and constructing a built-in.

HOW TO ADD

A WINDOW

Putting in a new window may not be as hard a job as you thought

Outside appearance and interior workshop area was enhanced considerably by the addition of this large window. Handy shelf on the extended bay was located over bench.

Our example shows how the job was done in a garage with uncovered walls. The procedure would be just about the same in a finished room; only difference—you have to do some "searching" to determine existing stud locations.

Best bet is to use chalk to draw the window outline on the wall. Tap with a hammer under this area until a solid sound indicates a stud. Studs to either side of this one will be located on 16" centers. Thus you can mark to both sides of the window plan to give you an X-ray view of what is behind the wall covering.

PLAN CAREFULLY

Before piercing the wall, check the drawing that shows the rough framing required for the new window. This consists of two *outline studs*, two *filler studs*, doubled *sill* and the *header*. Plan the installation for a minimum disruption of the existing structure. It might be possible to use two, or at least one, of the existing studs as the outline members. Of course, you should have the new window on hand—a label on the glass will tell you the exact size of the opening required. This will run vertically between header and sill and horizontally between filler studs.

REMOVE PANELS

If the wall is covered with panels (wood or plaster), it's a good idea to locate the joints and to remove entire panels even if it means uncovering more of the wall than is necessary. This actually makes the job easier than trying to cut through and will also simplify recovering since you will be able to start from joint places already there. If the wall is real plaster, remove—conversely—as little as you need to do the job, working from the center of the studs that represent extreme points.

116

It won't be too difficult to remove wood paneling so it can be reused. This can even be accomplished with a plaster panel but an authentic plaster covering must be destroyed and replaced.

After the wall is uncovered, cut the new header to length—as long as the window opening calls for *plus* twice the filler stud thickness. Assuming that you are repositioning studs for outline positions, it might be more convenient to make a sub-assembly of the rough framing of the floor. Assemble outliners, fillers, header and sill and then tilt into position and secure by toe-nailing to existing framework. If you work on the floor, use a square to check the window opening. If you work on the vertical, use a level to check *all* new pieces for vertical and horizontal alignment.

CUTTING THROUGH OUTER WALL

Now you have the complete rough framing for the new window and can cut through the outer wall. Start this by drilling a hole in each corner and then drawing lines on the outside wall to connect. You can cut through with a hand saw, a portable cut-off saw or even a saber saw. A builder might use a bayonet saw.

How you proceed from here depends on the window. Most metal sash windows are simply nailed on from the outside after plenty of caulking has been plastered on the flange. Outside framing fits in a groove provided for it, or it is set over the joint where the house siding meets the window. On some jobs, the siding butts against the window frame. But in any case a good weather seal is in order so don't neglect the caulking. Be especially careful at the top where rain water may run down. Usually, building paper is brought down and curled under the frame or the siding (whichever it may be) where it meets the window. Any excess paper is trimmed off with a sharp knife after the job is finished.

For a fixed glass window, you install a finish sill and inside stops that are wide enough to be flush with the finish wall. Glass is caulked between these stops and outside stops and then inside and outside frames are added.

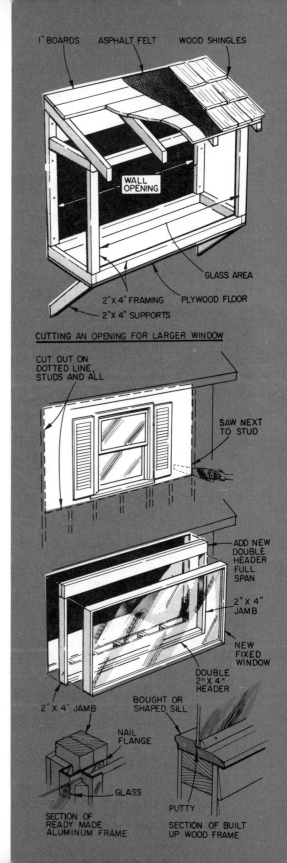

1" BOARDS ASPHALT FELT WOOD SHINGLES

WALL OPENING

GLASS AREA

2"x4" FRAMING PLYWOOD FLOOR

2"x4" SUPPORTS

CUTTING AN OPENING FOR LARGER WINDOW

CUT OUT ON DOTTED LINE, STUDS AND ALL

SAW NEXT TO STUD

ADD NEW DOUBLE HEADER FULL SPAN

2"x4" JAMB

NEW FIXED WINDOW

DOUBLE 2"x4" HEADER

2"x4" JAMB

BOUGHT OR SHAPED SILL

NAIL FLANGE

GLASS

SECTION OF READY MADE ALUMINUM FRAME

PUTTY

SECTION OF BUILT UP WOOD FRAME

BUILDING
FENCES

Plan for an ultimate goal so that what you put in will fit scheme

There are some factors to consider before you actually start construction. Your best bet, in order to preview what your lot will look like ultimately, is to make a simple sketch showing a bird's-eye view of the lot *and* the house. When you do this, make a careful analysis of how the area around the house will be used. Indicate a utility area (where you keep garden tools, trash cans, wheelbarrow, etc.), the site for the patio, playground area or maybe just a spot for a sand box, doghouse location and that

Make openings in screen by cutting some boards shorter than others. Stringer material used to frame openings for plaque. Good idea for inserting a planter box too.

sort of thing. Also indicate traffic patterns—how do you get from the garage to the house? What path do you use to get to the utility area? Are you going to house a trailer or a boat?

Based on this sketch you can make decisions about the *type* of fence you are going to put around each area. Also, you can now do the job in stages knowing it will end up being what you want.

THE BOARD FENCE

The most common type of privacy fence and one that is fairly easy and fast to construct is the board fence. This is usually purchased by the lineal foot and includes posts, stringers and boards. Most times the materials are figures with posts on 8-ft. centers. This is a good fence and serves many purposes but it is not the answer to all situations. For one thing the boards hide the framework on one side only. So you have to decide who

Patio screen made of economical board-fence material is deluxe job due to extra effort. Fence boards were cut to length and then inserted in dadoes cut in stringers and posts.

gets the "pretty" side; usually the decision is made in favor of the neighbors, which of course also makes the good side visible to anyone looking at your home.

BOARD-FENCE DESIGN

You *can* use this utility-type fence material to make a design that is the same on both sides. You can put solid sheathing on each side—which requires twice as many boards—or you can accomplish something similar without additional material by alternating board placement on opposite sides of the framing. The lead photo in this section shows a patio screen from board-fence material but one which has a deluxe feeling because of some imagination and a little extra work.

The idea, basically, is to cut dadoes in the posts and stringers to take the boards. Board edges are rabbeted so they overlap when installed. Thus the boards are framed by posts and stringers and there is no possibility of a gap between them. Also, the fence appears the same on each side. In some sections, openings were provided (as the sketches show) so plaques or flower boxes could be installed. The whole makes the patio much more pleasant than a solid board fence would have done.

Other ways to dress up a board fence are these: Dress the board-tops to make a more pleasant or intriguing line. Rip the boards to correct width and use them louvre-like *between* the stringers. Alternate board placement between posts. If you design post-spacing to suit board length, you can place one section of boards vertically, the next horizontally. Cut two-inch boards, from the same material and use them as batts to cover the joint between boards. This will give you nice shadow lines as the Sun moves.

The point is that you can do much with the economy materials by playing with them on paper before going to work.

119

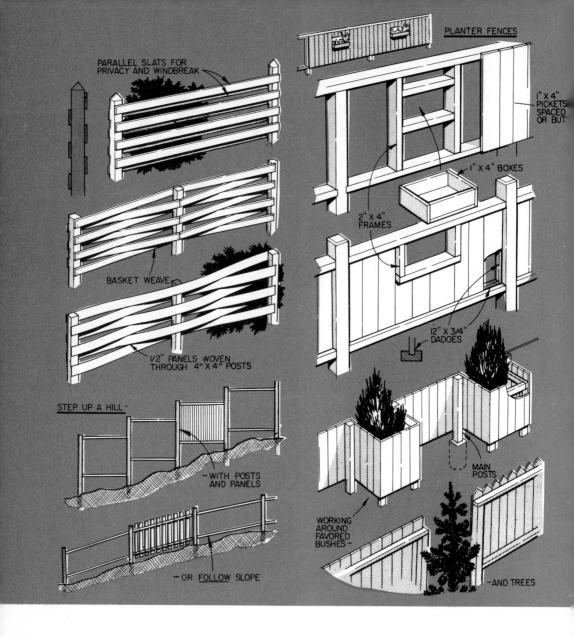

PARALLEL SLATS FOR
PRIVACY AND WINDBREAK

PLANTER FENCES

1" X 4"
PICKETS
SPACED
OR BUT

1" X 4" BOXES

2" X 4"
FRAMES

BASKET WEAVE

1/2" PANELS WOVEN
THROUGH 4" X 4" POSTS

2" X 3/4"
DADOES

STEP UP A HILL

– WITH POSTS
AND PANELS

MAIN
POSTS

– OR FOLLOW SLOPE

WORKING
AROUND
FAVORED
BUSHES –

–AND TREES

VARIETY OF MATERIALS

A list of materials good for fence-building would include just about anything you can name. Plywoods, hardboards, house siding materials, corrugated aluminum or fiberglas are among the most common. The sketches show various ideas for utilizing these and similar materials.

THE FENCE POSTS

Whatever your fence design, remember that the whole thing is supported by posts. These are made of 4×4" stock set a minimum of 18" in the ground. The ideal installation is in a hole 24" deep so you can provide a sub-base of gravel for water drainage. A good diameter for the hole is 6". This is rather difficult to ac-

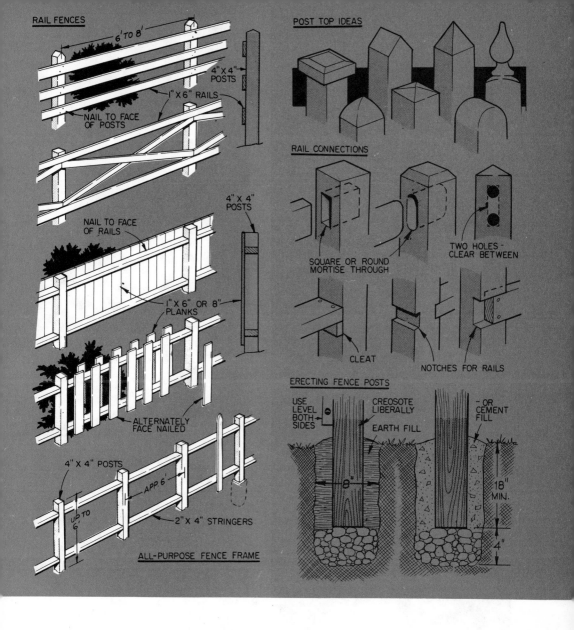

RAIL FENCES

6' TO 8'

4" X 4" POSTS

1" X 6" RAILS

NAIL TO FACE OF POSTS

NAIL TO FACE OF RAILS

4" X 4" POSTS

1" X 6" OR 8" PLANKS

ALTERNATELY FACE NAILED

4" X 4" POSTS

APP. 6'

UP TO 6'

2" X 4" STRINGERS

ALL-PURPOSE FENCE FRAME

POST TOP IDEAS

RAIL CONNECTIONS

SQUARE OR ROUND MORTISE THROUGH

TWO HOLES - CLEAR BETWEEN

CLEAT

NOTCHES FOR RAILS

ERECTING FENCE POSTS

USE LEVEL BOTH SIDES

CREOSOTE LIBERALLY

EARTH FILL

8"

- OR CEMENT FILL

18" MIN.

4"

complish with even the smallest shovel so it's a good idea to buy or rent a clam-shell digger. This is a two-blade, two-handled affair that is forced into the ground. Then the handles are spread so the blades grip the loosened soil for re-moval. This gives you control over hole-diameter. Another version has a cup-like blade with a cutting tip. You turn this as if you were man-powering an auger. The cup collects the loosened soil as you turn. The clam design is good for soft soil—the other is better for tougher con-ditions. Better than either, when you have a lot of holes to dig, is a post-hole digger which is a gasoline-powered auger. This is usually a two-man tool and will have to be rented but it makes form-ing any number holes of similar diameter a snap.

Resawn cedar boards used to sheath high fence. Note how good garden work destroys the closed-in feeling. When fences are tall use three stringers instead of only two.

For something fancy in the way of a rail fence, put the rails through the posts instead of surface-nailing them. Sketch shows you how to form the holes in the posts.

You can buy pre-fab fences. This one comes in pre-assembled section. After installing
posts, set panels between. You can Mix-or-Match the panels to satisfy your own taste.

There are good reasons for keeping the hole diameter to the recommended six inches. For one thing, it's wise to keep as much firm soil around the area as possible. For another, it minimizes the amount of concrete you need to set the post. It's also possible to call in professionals to supply the posts and the installation. They charge so much per post and it's possible, in this case, that it might be worth the cost. No harm to get a bid on it anyway.

IN GENERAL

Dig post holes after you have stretched a line to indicate the fence and have placed stakes to mark post locations.

Don't try to get correct post height by controlling post-hole depth. Instead, after posts are installed, stretch a line to indicate height and cut each post accordingly. Don't do any hammering on the posts until after the concrete has set. When setting posts, use a level on two adjacent faces to gauge verticalness. If necessary, use temporary braces.

Always work with galvanized or aluminum nails.

Be sure you know your property lines when you are building a perimeter fence. It would be a little frustrating to discover that the fence you built might have to come down.

Check local codes before building. Chances are there will be more restrictions on fence height at the front of the lot than at the rear. Codes will also tell you about set-back from the street.

It's a good idea to consult with your neighbor when building a fence on a common boundary. It will lead to better relations and maybe to sharing the cost and labor.

Build for the job the fence must do. If you own a sizeable chunk of property and neighbors are far away, a simple rail fence might do. This type will also do to contain horses. For complete privacy you want solid sheathing but this can be dressed up for better appearance. When you go for solid sheathing be sure you won't cut out any breezes that you might enjoy on warm days. A louvred fence would be better and also provide privacy. Think of a patio screen as being the wall of an outside room; you don't want to feel like you are sitting in a crate. Inside fences are screens, so don't forget to check prevailing wind direction. You may want to keep it out or let it in.

Most times, the reason for a boundary fence is to provide privacy. Just remember you have to live with and look at what you build to keep the world out. It's also wise to consider that a good job will increase property value—a poor one will down-grade it.

DECKS AND FOOTINGS

Most decks are built over sloping land, but can be flat on ground

A deck actually seems like an outdoor room, a lovely place to relax. Building out over a slope is practical. Outside edge is 7-ft. off ground. Hole for tree adds interest.

A deck is an outdoor floor built almost as solidly as you would construct a floor for a house to sit on. If you check the chapter on house framing you will see that the understructure design for a home built over a crawl space is pretty much what you need for a deck. You don't need the perimeter foundation and you don't go through the subfloor and finish floor routines. So a deck is a much easier job.

Most decks involve the use of pre-cast concrete piers set on small, poured footings, 4×4 posts, 4×6 beams, and either 2×4 or 2×6 flooring boards. You can go heavier on the beams but the only justification for this would be a greater span between posts than the 4′ recommended for the above materials. It's also possible to set posts directly in the ground if you follow directly the instructions shown in one of the sketches. Also, posts can be set directly on concrete footings that you pour yourself and this may be the way to

go in areas with deep frost lines. Which method you choose may be a matter of preference or it may be a matter of necessity. Best to check local building codes if you are in doubt. These will be very specific about the best thing to do for your area.

FOOTING AND PIERS

Whatever the method, don't try to achieve deck levelness by making footing or pier height uniform. This is better accomplished by stretching a line across the piers (to represent the beams) and cutting each post to fit. Each footing however, *does* have to be horizontal so check with a level when you install the footing forms or when you set down the piers. The wisdom of working this way is obvious on uneven ground or when you are building on a slope.

When the deck ties in to the house, make the connection there by attaching

A deck does not have to be big affair. This little one adds much interest to this prefab and makes for very practical entryway. More attractive than wood or concrete steps.

Forms for footings should be at least 6-in. deep. Check with a level when installing. This will make following jobs easier. The height is not important; firm base soil is.

Concrete piers "glued" to footings with strong mortar mix or set directly in footing pour. Be sure to use level here, too. Batter boards set for string framing site.

All piers are in and the job is ready for the posts. Note: no attempt was made here to get pier tops same height. Deck levelness will be achieved through post length.

Floor boards go down over beams and nailed at right angles to them. Tongue-and-groove boards are used here to make solid flooring (slanted for drainage) for carpeting.

a header or plate to the house wall. This should be spiked or lag-screwed through the wall into studs. The beams can hang from this on metal brackets you can buy or, as the sketches show, the beam ends can be notched to fit over a ribbon that is secured to the plate. The ready-made bracket is the easiest way to go.

POSTS ON PIERS

Posts are tied to piers by toe-nailing into the wooden block that is part of the pier. If you choose to — or must pour your own, consider the use of metal brackets that are set directly in the concrete. The post sits between the two

126

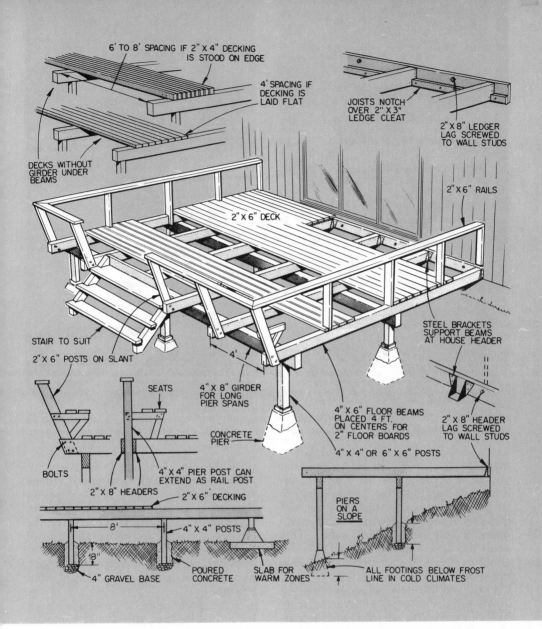

6' TO 8' SPACING IF 2" X 4" DECKING IS STOOD ON EDGE

4' SPACING IF DECKING IS LAID FLAT

DECKS WITHOUT GIRDER UNDER BEAMS

JOISTS NOTCH OVER 2" X 3" LEDGE CLEAT

2" X 8" LEDGER LAG SCREWED TO WALL STUDS

2" X 6" RAILS

2" X 6" DECK

STEEL BRACKETS SUPPORT BEAMS AT HOUSE HEADER

STAIR TO SUIT

2" X 6" POSTS ON SLANT

SEATS

4" X 8" GIRDER FOR LONG PIER SPANS

CONCRETE PIER

4" X 6" FLOOR BEAMS PLACED 4 FT. ON CENTERS FOR 2" FLOOR BOARDS

4" X 4" OR 6" X 6" POSTS

2" X 8" HEADER LAG SCREWED TO WALL STUDS

BOLTS

4" X 4" PIER POST CAN EXTEND AS RAIL POST

2" X 8" HEADERS

2" X 6" DECKING

PIERS ON A SLOPE

8'

4" X 4" POSTS

18"

4" GRAVEL BASE

POURED CONCRETE

SLAB FOR WARM ZONES

ALL FOOTINGS BELOW FROST LINE IN COLD CLIMATES

flanges of the bracket that extend above the concrete. You just nail through predrilled holes in the bracket into the post.

Beams are tied to post-tops, again by toe-nailing, or by cutting splice plates from 3/4" stock and surface-nailing into both beam and post. The plates should be as wide as post at least 12" long.

DECK FLOORING

Flooring is secured by surface-nailing into the beams. Most times, a space is allowed between the boards so you won't have a water problem when it rains or when you hose down the deck or sweep it. This space can be as much as a half-

Ideal setting for deck. Note outward curving rail, not too difficult to accomplish if you own a bandsaw or a heavy-duty saber saw. Douglas fir used to build this flooring.

inch but the small amount of material you save by doing this isn't wise economy. For one thing the thin heels worn by women can easily slip into that half inch gap. Better to keep the gap to a minimum—a 16d nail is a good gauge. This gives the appearance of a solid floor, is about as practical and still accomplishes the water-elimination job.

Don't be careless when hammering down the deck boards. Drive the nails to about the surface of the board then set them just a fraction with a flat-face punch. Don't use finishing nails—16d galvanized common nails are what you want. Drive two nails through each flooring board at each beam crossing. Don't try to set them just by using the hammer. This will leave dents that will spoil the appearance of the floor regardless of whether you end up with a natural finish, stain or paint.

Redwood, because it stands up so well outdoors and because it weathers to some beautiful tints, is a popular deck material but there are many other species that are fine for the job—Douglas fir, cedar, pine, larch. Make a choice on the effects you want and the availability of materials in your area. Some lumber does best outdoors when painted, others, like redwood and cedar, do okay without a finish.

Some decks can be ground-level affairs. In this case you would eliminate footings, piers and posts. Beams, if they make direct contact with the soil, should be of redwood, and also should be soaked in a preservative. Best way to do this is—dig a trench in the ground long enough and wide enough to take a beam. Line the trench with sheet plastic and then fill it with a preservative. Then you can soak the beams before installing them. This does a better job than application by brush.

INDEX